Reality Lost

Vincent F. Hendricks
Mads Vestergaard

Reality Lost

Markets of Attention,
Misinformation and
Manipulation

 Springer Open

Vincent F. Hendricks
Center for Information and
Bubble Studies (CIBS)
University of Copenhagen
Copenhagen S, Denmark

Mads Vestergaard
Center for Information and
Bubble Studies (CIBS)
University of Copenhagen
Copenhagen S, Denmark

Translated from the Danish by Sara Høyrup / Hoyrup.biz - English
language copy editing by Vincent F. Hendricks

ISBN 978-3-030-00812-3   ISBN 978-3-030-00813-0   (eBook)
https://doi.org/10.1007/978-3-030-00813-0

Library of Congress Control Number: 2018955285

Front cover design by Michel Winckler-Krog

This Springer imprint is published by the registered company Springer
Nature Switzerland AG
The registered company address is: Gewerbestrasse 11, 6330 Cham, Switzerland

Prelude

Knowledge is power. To know the nuts and bolts of the world is vital in changing it to the better—or to prevent it from getting worse. Knowledge may emancipate. Misinformation and manipulation impair self-determination and autonomy—individually and collectively. Without sound information as bedrock for formation of political opinion, decision-making, and action, individual agency and political sovereignty of the people are crippled. To know how and why misinformation works, being informed of the structural conditions fueling its circulation shapes resilience against manipulation. Technology does not determine history. Humans do. It is up to us to prevent the digital age turning into disruption of democracy and freedom.

Reality Lost: Markets of Attention, Misinformation and Manipulation is open access and free to download.

Copenhagen S, Denmark   Vincent F. Hendricks
July 2017   Mads Vestergaard

Acknowledgments

We would like to thank Center for Information and Bubble Studies (CIBS) at the University of Copenhagen and in particular our team Mikkel Birkegaard, Thomas Bolander, David Budtz Pedersen, Ditte Dyrgaard, Benjamin Rud Elberth, Robin Engelhardt, Paolo Galeazzi, Kathrine Elmose Jørgensen, Hanna van Lee, Laurs Leth, Camilla Mehlsen, Silas L. Marker, Esther Kjeldahl Michelsen, our invaluable center administrator Maj Riis Poulsen, Jan Lundorff Rasmussen, Rasmus K. Rendsvig, Joachim Wiewiura, Mikkel Vinther, Christiern Santos Rasmussen, Michael Hansen, Anders Rahbek, Peter Norman Sørensen, Frederik Stjernfeldt, and the CIBS Scientific Oversight Committee consisting of Adam Brandenburger (NYU), Richard Bradley (LSE), Robert Becker (IU Bloomington), and Nina Smith (AU). A special thanks to Jerome L. Coben for his pertinent comments and constructive criticism on earlier drafts of *Reality Lost*.

Our gratitude is also expressed to The Carlsberg Foundation, and in particular Chairman Flemming Besenbacher, for providing the grant establishing CIBS together with the Faculty of Humanities of the University of Copenhagen. Our thanks are also extended to the Tryg Foundation for funding the D.U.D.E., the digital education program, which has continuously also educated us while putting this book together.

Vincent would like to thank Thomas Buch-Andersen, Adam Holm, Bjarke Malmstrøm Jensen, Poul Madsen, Lise Nørgaard, Annette Møller, Teit Molter, Søren Pind, Frederik Preisler, Simon Vesth — and his family.

Mads would to thank Caroline Hjorth Saabye for her support in private and in profession. Also, he would like to thank Uffe Vestergaard and Sanne Andersen and Martin Vestergaard, Magnus Vestergaard and Sofie Vestergaard, Lis Vestergaard and the whole Vestergaard and Andersen family together with Linette Larsen, Thomas Vass, Lasse Lauridsen, Søren Fisker, Bjørn Rosenberg, Dennis Kristensen, Thomas Christensen, Michel Winckler-Krog, Keith Kawahata, Miguel Paley, and Thilo Traber. A special thanks for Frank Lindvig. His thoughts are manifested in Chap. 5.

Finally we would like to thank our editor at Springer Nature, Ties Nijssen, for his enthusiasm pertaining to the project and his efficiency in getting it out there.

Introduction

On January 4, 2017, a meager and obscure news outlet called Donbass News Agency ran a story reporting that the USA was on the verge of sending 3600 tanks to Europe as part of "The NATO war preparation against Russia" (Fig. 1).[1]

Within days, the story went viral. It appeared on several media in the USA, Canada, and Europe, and it was shared 40,000 times, translated into Norwegian, and quoted by the official Russian news agency RIA Novosti. It wallowed in attention, particularly from the Russian press.[2]

This parcel of information was actually misinformation. True, the USA was planning to increase its forces in Europe, but not in tanks and certainly not in those *numbers* claimed by Donbass News Agency. Such numbers would have made the presence of American tanks 20 times larger than it actually was. This is how misinformation oftentimes works: It is not entirely false, but rather a synthesized pill of downright false or misleading information mixed up and sugar coated with a grain of truth making it easier to swallow uncritically.

Only 2 days prior to the second round of the French Presidential Election in May 2017, candidate Emmanuel

---

[1] Donbass News Agency, January 4, 2017. "US sends 3.600 tanks against Russia – Massive NATO deployment on the way," verified April 4, 2017: https://dninews.com/article/us-sends-3600-tanks-against-russia-massive-nato-deployment-underway

[2] Digital Forensic Lab (2017): "Three Thousand Tanks" (January 12, 2017), verified April 4, 2017: https://medium.com/@DFRLab/three-thousand-fake-tanks-575410c4f64d

FIG. 1. The news story from Donbass News Agency that went around the world.

Macron's campaign suffered a massive hacking attack in which nine gigabytes' worth of emails and other internal communication and memos were leaked and spread on social media. According to the Macron campaign, the leak was a combination of authentic material mixed with forged documents and fabrications, with the intent to sow "doubt and misinformation."[3] A press release from the Macron campaign *En Marche!* confirmed the leak and stated: "The seriousness of this incident is without a doubt, and we cannot tolerate the endangering of vital democratic interests."[4]

As early as 2013, the World Economic Forum announced:

> The global risk of *massive digital misinformation* sits at the centre of a constellation of technological and geopolitical risks ranging

---

[3] Auchard, E. & Felix, B. (2017): "French candidate Macron claims massive hack as emails leaked," *Reuters*, May 6, 2017, verified May 7, 2017: http://www.reuters.com/article/us-france-election-macron-leaks-idUSKBN1812AZ

[4] Chung, A. (2017): "Macron team condemns 'massive cyberattack' ahead of French presidential election," *Sky News*, May 6, 2017, verified May 7, 2017: http://news.sky.com/story/macron-team-target-of-massive-cyber-attack-10865052

> from *terrorism* to *cyber attacks* and *the failure of global governance.*[5]

In the digital age, misinformation has become a global challenge joining the family of anthropogenic climate change, increasing economic inequality, water supply shortage, global health problems, and a range of other urgent problems. Digital misinformation is not only to be blamed on foreign powers' attempts at interference. The challenge cannot be met simply by pointing fingers at villains who create — or pay to have created — the fallacious fables finding their way to the news. Villains indeed exist, but if attention is only restricted to shady *actors*, the *structural* conditions that pave the way for misinformation are not acknowledged. That would make it difficult or even impossible to curb the level of misinformation and the damage it does.

The evidence of Russian attempts to influence the outcome of Brexit and the US Presidential Election in 2016[6] has revealed how vulnerable the current digital environment of information actually is. In order to protect against attacks in the guise of (mis)information and against external influences that might have an impact on political opinion formation and democratic elections, it is imperative to fully understand the structure of this vulnerability and what has created it.

The traditional news media are a key part of the puzzle. We expect the written and spoken press to act as a watchdog and an entity exercising checks-and-balances: the bouncers of the public sphere and truth's keeper. How the press acts and reacts to potential misinformation is key to the efficiency and damaging capacity of the very misinformation in question.

If ethical guidelines are pushed to second base in how news is covered, even established news outlets may actually

---

[5] WEF (2013): "Digital Wildfires in a Hyperconnected World," verified May 5, 2017: http://reports.weforum.org/global-risks-2013/risk-case-1/digital-wildfires-in-a-hyperconnected-world/

[6] Lomas, N. (2017). "Study: Russian Twitter bots sent 45k Brexit tweets close to vote," *TC*, 15.11.2017, verified 26.06.2017: https://techcrunch.com/2017/11/15/study-russian-twitter-bots-sent-45k-brexit-tweets-close-to-vote/

contribute to the misinformation, rather than expose and avert it. The existence or nonexistence of ethical principles and the nature of the intentions of journalists and media institutions are only some of the pieces in the grander puzzle. They do not make up the full picture, and they cannot be the solution in and of themselves.

Media and journalists operate in an environment that sets the stage for news coverage and reporting. In order to survive in any given environment, it is necessary to adapt to its conditions. This goes for those in the media industry as well. The economic conditions for journalism influence its quality and corollaries. The public service call of media outlets to inform citizens, the business models, and market conditions also make for a difference. If the news market is fully commercialized and completely dependent on advertisers whose only criterion for quality is integers of ears and eyeballs, entertainment value, conflict, sensation, if-it-bleeds-it-leads may well become the decisive news criteria. Here is a possible beeline to misinformation, populism, and political manipulation (Fig. 2).

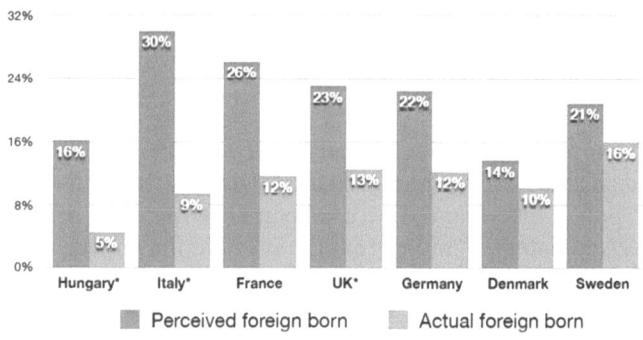

Survey data from ESS (2014) and Ipsos MORI* (2013). Foreign-born population data from Eurostat (2014 data).

Fig. 2. The perception among citizens in seven EU countries pertaining to the percentage of their country's total population born abroad, compared to the real numbers. While the difference between perception and reality is only 4% in Denmark at one end of the scale, it is at a full 21% in Italy at the other end. Source: (Flynn et al. 2017).

Traditionally, tyrants' strategy has been to keep the people's information level at an absolute minimum as means of exerting power. Through censorship and punishment, oppressors could deprive people of the sources of information considered problematic. Some rulers still have a jab at this strategy today, and the struggle *for* freedom of speech is a struggle *against* this ruling strategy.

However, in the age of information, a similar propagandistic effect may be obtained by rather *drowning* than depriving citizens, voters, and media in misinformation and noise. This realized without censorship and suppression of freedom of speech. It may turn out quite effective if the level of trust in the press as watchdog is low—for good or bad reasons. Freedom of speech is not a bulwark on its own against such tactics of information control. What may stem the tide without limiting freedom of speech and accordingly undermine freedom, enlightenment, and democracy? In order to approach an answer to this complex, yet very urgent question, a prerequisite is understanding the technological and commercial, as well as psychological, conditions that make misinformation so potentially potent, profitable, and perilous.

Digitalization of information and media content coupled with the infrastructure of the Internet makes the production and proliferation of misinformation possible at a whole new level. The market for media products, which the Internet has created, grants favorable propaganda possibilities for vested political and economic interests. The information ecosystem and economy offer strong financial incentives to produce and spread distorted stories, rumors, and fake news: They are highly contagious, seize attention, and generate clicks. Due to automated advertising systems, clicks on the Internet may be traded directly for cash. But the truthfulness of stories generating the clicks is of little to no consequence: A click is a click.

In a worst-case scenario where the amount of circulated and accepted misinformation running wild, a democratic polity may approach a post-factual state.

> A democracy is in a **post-factual state** when politically opportune but factually misleading narratives form the basis for political debate, decision, and legislation.

Should it come to such a state, facts and knowledge are devaluated. Political and practical ability to address and solve social and global challenges diminish. Not only that, some 5 years after establishing digital misinformation as a global risk, the World Economic Forum now concludes that democratic governance itself is threatened by misinformation.[7]

This book provides the nuts and bolts of an explanatory framework for the devolving of democracy to a post-factual state. The first five chapters cover the building blocks, conceptual scaffold and conditions for emergence of a post-factual state, while the last chapter explains what post-factual democracy is and why ending there is to be prevented.

In the digital age, there is no shortage of information. Consuming information comes at a price—we pay with our *attention*. Our attention is the portal to our consciousness, which makes our attention valuable to anyone with a message, a news story, or a product to sell. When we bestow our attention on something specific (such as following a Facebook debate on the supposed threat posed by immigration; catching yet another tweet from Donald J. Trump on Barack Obama's supposed bugging of Trump Tower and the FBI undercover involvement during campaign; watching a rumor claiming that Chancellor Angela Merkel runs the German media house ZDF and endorses child brides; or that Anthony Bourdain of CNN was killed by Clinton operatives while Chelsea Clinton admits that #Pizzagate was real), it deflects our attention away from other matters.

There are only so many hours in the day; our attention is a scarce resource. That is why everyone is fighting so hard to get a piece of it. The outcome of this struggle often decides what sets the agenda in the news stream and in politics, whether online or off-line. The attention economy and the

---

[7] World Economic Forum: *Global Risks Report* 2017: 24. Verified June 11, 2017: https://www.weforum.org/reports/the-global-risks-report-2017

market for news and political messages are the themes of Chaps. 1 and 2, respectively.

As in the monetary economy, it is also possible to *speculate* and create political bubbles void of substance in the market for news and political messages. Those are the focal points of Chap. 3.

Chapter 4 aligns and orders the quality of information; from misinformation to disinformation; from truth over misrepresentations and doctored statements to unconfirmed rumors and concealment ending up with downright lies, bullshit, and fake news.

Chapter 5 covers a number of the psychological and socio-psychological mechanisms that make us resistant or even *immune* to facts and render misinformation effectual. The same dynamics make populism and simple "us-versus-them" narratives an apt political strategy by smearing the media, the elite, the Rust Belt, Brussels, the clueless, the uninformed, Washington, the rich, the poor, and the foreigners, and by discrediting political opponents. If "the others" are perceived as nothing short of an *enemy*, truth is the first victim in a war between factions marked by distrust. This may allow conspiracy theories feeding of distrust—as well of fueling it—to live long and prosper to the expense of civil, factually sound, and reasoned democratic deliberation.

The final Chap. 6 chronicles the post-factual democracy and explains why it is not very democratic at all and certainly worth combating, just like its limit point on the other end of teeter-tooter: technocracy.

The Epilogue (Chap. 7) reviews dystopian perspectives of a novel of digital totalitarianism made possible by the digital revolution. The dream and promise of digital emancipation may turn into its opposite. Under the guise of "enhanced user experiences," the digital technology and the markets thus created may pave way for a surveillance society in which new effective methods of control undermine autonomy and self-determination. A first step in preventing this from happening is to acknowledge the conditions, mechanisms, and motivations driving the markets of attention, misinformation, and manipulation.

Contents

About the Authors

Vincent F. Hendricks is Professor of Formal Philosophy at the University of Copenhagen. He is Director of the Center for Information and Bubble Studies (CIBS) sponsored by the Carlsberg Foundation and has been awarded a number of prizes for his research, among them the Elite Research Prize by the Danish Ministry of Science, Technology, and Innovation, the Roskilde Festival Elite Research Prize, Choice Magazine Outstanding Academic Title Award, and the Rosenkjær Prize. He was Editor-in-Chief of *Synthese*: an international journal for epistemology, methodology, and philosophy of science between 2005 and 2015.

Mads Vestergaard (b. 1979) is PhD Fellow at Center for Information and Bubble Studies (CIBS), University of Copenhagen. He holds a M.A. in philosophy, and debates and communicates philosophy and social science in written press, radio and television. He has also applied philosophy in more untraditional contexts as art projects and satire.

The original version of this book was revised to include the following text in the page iv, "Translated from the Danish by Sara Høyrup / Hoyrup.biz - English language copy editing by Vincent F. Hendricks." The correction to this book can be found at https://doi.org/10.1007/978-3-030-00813-0_8

Chapter 1
The Attention Economy

1.1 The Information Society

According to legend, Abraham Lincoln was willing to walk several miles in order to borrow a book while growing up in Indiana during the early nineteenth century. "My best friend is the man who'll get me a book I ain't read," young Lincoln is reported to have said.[1] Literature was scarce, difficult to access, and precious. Not only literature but information in general was hard to come by. Whether news from afar, new knowledge and insight, or simple entertainment, it usually took effort and came at considerable expense to get hold of information.

Just a few years ago, information was much more difficult to get hold of than it is today. Being well-informed depended on subscribing to a newspaper, heading out to buy one, or going to the library. Digitization and information technology have changed all this. Today, a smartphone is enough: access to any information of choice, let it be news, politics, or scientific results; literature, entertainment, or gossip; or endearing baby pictures or cute cat videos. Never before has so much information been so easily accessible.

---

[1] Holleran, A. (2008): "Such a Rough Diamond of a Man," *New York Times*, July 11, 2008. Verified June 27, 2018: http://www.nytimes.com/2008/11/09/books/review/Holleran-t.html

© The Author(s) 2019
V. F. Hendricks, M. Vestergaard, *Reality Lost*,
https://doi.org/10.1007/978-3-030-00813-0_1

The hallmark of the information age is not that we are all in continuous pursuit of precious information hard to access, but the other way around: The information age offers so much information that drowning in it, or chocking on it, is the risk. The vast offer of freely available information online has made the value of information drop steeply. People grown up with the Internet expect to get their information for free and refuse to pay for newspapers, books, or entertainment products. Not too many people would be willing to walk for miles to get hold of a book in this day and age.

1.2 The Price of Information

The easy access to overwhelming amounts of information, and the fact that often you don't have to pay money for it, doesn't mean that information comes for *free*; to receive information, we pay attention. You may have access to loads information, but in order to take it in, process it, and possibly act on it, you spend your attention on it. Project Gutenberg has made more than 53,000 books freely accessible online. If you read a book a day, it will take you 145 years to get through a library that size. If you prefer video, 400 h are uploaded to YouTube every minute. The challenge today is not to find something to read or information to pay attention to; it is to find the time to read or look at the material at your disposal.

With information in abundance comes an attention deficit. As early as 1971, Nobel Prize Laureate in economics Herbert Simon prophetically said about the information age to come:

> …in an information-rich world, the wealth of information means a dearth of something else: a scarcity of whatever it is that information consumes. What information consumes is rather obvious: it consumes the attention of its recipients.[2]

---

[2] Simon (1971: pp. 40–41).

The fact that information consumes attention makes attention a valuable resource. The information taken in is the basis of our experience and knowledge and deliberation, decision, and action.

Attention is a curious resource compared to economical means since it is more equitably distributed. Surely, some people can concentrate longer and more intensively than others. All the same, there are only marginal differences in the amount of attention each of us can spend. Attention cannot be accumulated and saved like money for a rainy day. In our waking hours, we constantly spend our attention: We are always attentive to *something*. A common feature of both attention and money is that spending it on one thing excludes spending it on another.

1.3   The Scarcity of Attention

Philosopher and psychologist William James (1842–1910) has described attention in a famous quote from 1890:

> [Attention] …is the taking possession by the mind, in clear and vivid form, of one out of what seem several simultaneously possible objects or trains of thought … . It implies withdrawal from some things in order to deal effectively with others.[3]

In order to efficiently take in, process, and act on information, we need to focus on one thing at a time. This has been confirmed in recent cognitive research: Even if we may sometimes multitask and pay attention to several things at once, such as talking on the phone while cooking, it generally makes us slower and more prone to making mistakes. Quality wanes when we split our attention rather than focus on one single item or activity (Sternberg and Sternberg 2012) (Fig. 1.1).

---

[3] James, W. (1890): *The Principles of Psychology*, Chapter XI: Attention. *Classics in the History of Psychology*, Green, C.D. (ed.). Verified May 31, 2017: http://psychclassics.yorku.ca/James/Principles/prin11.htm

Fig. 1.1. Multitasking makes the quality of one's attention wane, as witnessed in our reflexes and the scores of information absorbed.

Time becomes a decisive factor when attention really only may be paid to one thing at a time. But time itself is fixed and limited: No matter how much we try to get organized and optimize with to-do lists, there are only 24 h in a day. We have a limited attention capacity (Kahneman 1973). This produces an upper bound for how much each of us may pay attention to, and therefore how much information we can take in and process on a daily basis. This makes the *selection* of information and *allocation* of attention of crucial importance.

Economics has been described as the "study of the allocation by individuals and societies of scarce resources" (Samuelson and Nordhaus 2010). When attention is viewed as a scarce resource, it creates the basis for studying the information age as an *attention economy*.

1.4   Information Sources

In order to obtain information about matters beyond our immediate environment, the media is needed as vehicles and presenters of information. This allots a very central role to the news media. Information is to a large extent received via channels created by the media. Therefore, the media's reliability as an information source is key to how well-informed, misinformed, or disinformed for that matter we are. If you do not pay attention to news or politics, but allocate your attention to entertainment, it should come as no surprise that you are hardly as informed about politics, as you would have been had you paid attention to it. And if attention is paid to unreliable sources and untrustworthy information, there is a greater risk of being deluded and duped. If attention is systematically spent on conspiratorial YouTube videos and political propaganda sites, no surprise either it invariably will color your perception of reality. A significant consumption of false claims, unconfirmed rumors "alternative facts," and fake news may cause you to lose your grip on the real world and relegate you to an alternative reality (Fig. 1.2).

When attention is consumed by information, information is the source of knowledge, and attention is a scarce resource, it is important to spend attention with care. This is easier said

FIG. 1.2. One-sided news diets may result in distorted ideas of reality.

than done. Many actors in the information market fight dirty to catch and harvest our attention.

1.5   The Market for Attention

Few people like being ignored, not seen nor heard by others. As individuals, we crave at least minimal dose of attention from other people and need it to thrive, both as children and adults. A lot of people just can't get enough, judging from the present-time celebrity and reality television culture. The pursuit of *fame* as reality star on TV or as a *micro celebrity* (or *micro influencer*) on social media may look like a pursuit of attention for the sake of attention itself (Marwick 2015) (Fig. 1.3).

Once in possession of people's attention, it may be transferred to others. If a stage performer points to one person in the audience, a large part of the audience's collective attention will be transferred from the performer to the happy fan. If you have people's attention, you can channel it to another person or product and monetize it. This is the principle in overt sponsorships and product placement alike. When the name of the firm goes on the player's T-shirt, or a media darling is paid to wear a brand visible to the cameras, the advertiser is purchasing into the audience's attention.

Marketing is intrinsically linked to attention harvesting. The aim of marketing is to influence *behavior*. Marketing is about persuading consumers into buying a certain product or voting for a specific candidate. No persuasion happens if no one is listening, reading of watching. Attention is the portal to people's minds and a necessary condition for all successful communication; from teaching and knowledge presentation to persuasion, seduction, and manipulation. This makes attention extremely valuable for everyone with something to sell. It is the main factor in all forms of marketing, branding, and advertising (Teixeira 2014).

Models of marketing qualify different levels of attention, ranging from no attention to partial attention (due to multi-

FIG. 1.3. Attention is pursued for its own sake, but may also be traded for sponsor and advertising revenue. Here is Kim Kardashian trying to "break the Internet," a metaphor for harvesting enormous amounts of online attention. (Spedding, E. (2016): "The man behind Kim Kardashian's Paper Magazine cover on how to break the Internet," *The Telegraph*, June 18, 2016, verified June 6, 2017: http://www.telegraph.co.uk/fashion/people/the-man-behind-kim-kardashians-paper-magazine-cover-on-how-to-br/).

Optimize Video Advertising for Your Audience's Attention Level

Should you engage, persuade, or a little of both?

CONTEXT	ATTENTION LEVEL	ADVERTISING STRATEGY
CINEMA	Full attention	*Focus on persuasion:* use mostly information
TELEVISION MULTITASKING	Partial attention (mostly to ad screen)	*Balance both goals:* entertainment and information
MOBILE MULTITASKING	Partial attention (mostly to second screen)	*Compete for attention:* entertain on one screen, inform on the other
PEER-TO-PEER SHARING	Lack of attention	*Critical to gain attention:* entertain to grab attention from few, inform over time

SOURCE THALES TEIXEIRA   ‹ HBR.ORG

FIG. 1.4. Strategies for catching or exploiting attention with a view to affecting behavior. (*Harvard Business Review* (2015), verified June 10, 2017: https://hbr.org/2015/10/when-people-pay-attention-to-video-ads-and-why).

tasking) to full attention (Fig. 1.4). The goal is to isolate the best marketing strategies given the attention already at the advertisers' disposal. If there is no attention, then attention needs to be seized; if there is only partial attention, it needs to be won over completely; and if someone's undivided attention is won over, it must be kept and used as efficiently as possible to persuade and affect behavior.

1.6   Attention Merchants

The intimate connection between attention, communication, and marketing forms the basis of an industry that Columbia Law School Professor Tim Wu has labelled *attention merchants* (Wu 2016). The basic business model is quite simple: harvest attention and resell it for marketing and advertising purposes.

Benjamin Day was the inventor of this business model and one of the brains behind the Penny Press in the 1830s. Back then, newspapers such as *The New York Times* and *The Wall Street Journal* cost six cents; they were luxury items for the privileged few. In 1833, Benjamin Day launched *The New York Sun* at one cent a paper, dumping the price.

The news criteria were also dumped. The only criterion for the stories was how many papers they could sell. Sensational, dramatic, and juicy crime copy was also popular back then. Material was picked up on a daily basis from the police departments and the courts. Crime sold newspapers. Benjamin Day was no journalist; he was a businessman. His newspaper took tabloid to whole new levels in order to achieve readership—for example, running a successful series in 1835 reporting on a new "scientific" discovery of a race of bat people inhabiting the Moon. Flavor was added to the story with claims of the libertine and promiscuous lifestyle of the bat people. Fake news is *not* a new invention (Fig. 1.5).

When the sole criterion for success is to sell as many papers as possible, truth is of little or no consequence, and a lot of papers had to be sold to make the project fly. The

FIG. 1.5.   Fake news from *The New York Sun* in 1835.

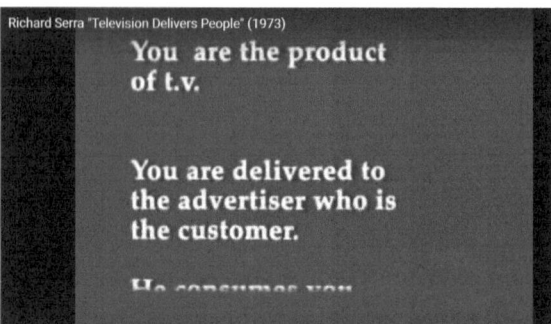

Fig. 1.6. Artist Richard Serra's minimalistic video work, "Television Delivers People," from 1973. (Richard Serra (1973): "TELEVISION DELIVERS PEOPLE," verified May 4, 2017: http://www2.nau.edu/~d-ctel/mediaPlayer/artPlayer/courses/ART300/pov1_ch1/transcript.htm).

papers were sold at less than the production costs. Selling a lot of papers in itself would have simply worsened the deficit. Indeed a bad business if the real customers were the readers paying a cent for the newspaper. In the attention merchant business model, however, the readers are actually the product sold to the real customers: the advertisers. The real customers for Benjamin Day were the companies who placed ads in *The New York Sun* to buy the attention of the readers.

The same business model is later being rehearsed on commercial TV competing for *eyeballs*. The viewers' attention is sold to the paying customers in the advertising industry. The industry in turn presents its messages and influences a broad public audience during commercial breaks. Ever more viewers, ever more attention is sold, and the higher the price for advertising seconds (Fig. 1.6).

From a business perspective, TV programs are merely means for selling what it is really all about, advertisements. The purpose of the programs on commercial TV is to make you watch more TV and stay on the channel. *Stay tuned*.

In the wake of the digital revolution, (business) history repeats itself. As the saying goes on online social networks and platforms; if you are not paying for the product, you *are* the product. If you perceive services such as Google and Facebook as truly free of charge, you have misunderstood the business model and your own role in it. The main default business model online is the attention merchant. Media researcher Douglas Rushkoff points out:

> Ask yourself who is paying for Facebook. Usually the people who are paying are the customers. Advertisers are the ones who are paying. If you don't know who the customer of the product you are using is, you don't know what the product is for. We are not the customers of Facebook, we are the product. Facebook is selling us to advertisers.[4]

The attention and data of the users are the items offered for sale to possibly third party. And similar to the casino, the more engagement by the users, the more social media stand to benefit. Like Robert de Niro says in *Casino* the movie: "In the casino, the cardinal rule is to keep them playing and keep them coming back. The longer they play, the more they lose. In the end, we get it all."

1.7 Data Collection

Corporations such as Facebook, Amazon, and Google collect enormous amounts of data about the online behavior of users. Together with masses of smaller players who also offer seemingly free products, not only do they sell user attention to advertising third parties; they sell a plethora of information about users. This goes for all the information shared when users fill out profiles, listing interests, age, gender, political affinity, relationship status, etc. Every piece of information given up has value when aggregated. This also applies to a heap of data constantly generated about our online behavior through cookies and other invisible tracking systems. Data

---

[4]Solon, O. (2011):"You are Facebook's product, not customer," *Wired*, September 21, 2011. Verified May 4, 2017: http://www.wired.co.uk/article/doug-rushkoff-hello-etsy

about everything, from searches and search patterns, visited pages, and engagement on social media to e-mail contacts and consumption patterns, are collected. Unless your phone came out of the ark, the same goes for your physical movements. If a child has a Hello Barbie doll, it collects and sends information back to the producer Mattel about what the child talks about, likes, and wishes for.[5] The collected data may be traded in a flourishing market where information on users and citizens is a valuable asset. Online surveillance and resale of the information generated by the surveillance is a growing industry. Already in 2012, the American data broker industry generated revenues ($156 billion) exceeding twice the amount the US government allocated to its whole intelligence budget.[6]

With data collection, online media and businesses have taken things a step further than earlier attention merchants. Not only do they sell the audience's attention; the collected information is used to target ads for each individual user so that the ads hits home pertaining to user needs, interests, and stances. In Facebook's own words:

> We want our advertising to be as relevant and interesting as the other information you find on our Services. With this in mind, we use all of the information we have about you to show you relevant ads.[7]

In less unequivocal terms, users are being monitored to the end of making economic profit by selling information about users along with their attention to third parties. Corporations like Facebook and Google secure their profit by means of a

---

[5] Marr, B. (2016): "Barbie Wants To Chat With Your Child—But Is Big Data Listening In?", *Forbes*, December 17, 2015. Verified June 12, 2017: https://www.forbes.com/sites/bernardmarr/2015/12/17/barbie-wants-to-chat-with-your-child-but-is-big-data-listening-in/#2b31020a2978

[6] Senator John D. Rockefeller IV (2013). "What Information Do Data Brokers Have on Consumers, and How Do They Use It?" December 18, 2013, Verified June 27, 2018: https://www.gpo.gov/fdsys/pkg/CHRG-113shrg95838/pdf/CHRG-113shrg95838.pdf

[7] *Facebook Data Policy*, verified 04.05.2017: https://www.facebook.com/privacy/explanation

business model that is based on surveillance (Taplin 2017). Surveillance provides information about the surveilled that may be (mis)used to persuade, trick, and manipulate more effectively.

1.8 Hit Them Where It Hurts

The amount of collected data combined with powerful computers make it possible to predict quite personal things that users otherwise would not share publicly. Even if you do not state your gender and age and where you live, all the other data points collected from your phone, your computers, your credit cards, etc. are enough for this basic information to be computed with accuracy. And this knowledge is worth a lot in marketing terms. It is much easier to persuade someone to do something or influence their behavior if you know them and know which buttons to push.

The company Target decided to compute whether women were pregnant, even if they had not given that information. That would be useful for marketing during the pregnancy: "We knew that if we could identify them in their second trimester, there's a good chance we could capture them for years …As soon as we get them buying diapers from us, they're going to start buying everything else too."[8] Target succeeded in this profiling endeavor. About a year after the onset of this pregnancy-targeted marketing campaign, a father turned up in one of the stores, upset that his 17-year-old daughter had received a pregnancy-related advertising e-mail. When the store manager spoke to the father over the phone later, it was the father who apologized; his daughter *was* actually pregnant.[9] It doesn't stop with pregnancy predictions. Big data's power of prediction may also establish your

---

[8] Duhigg, C. (2016): "How Companies Learn Your Secrets," *New York Times*, February 16, 2016. Verified June 12, 2017: http://www.nytimes.com/2012/02/19/magazine/shopping-habits.html?pagewanted=all&_r=1
[9] Ibid.

political views, religious beliefs, sexual orientation, and other very private, personal, but very useful information.

These data may be used for other purposes than showing you "relevant" advertising, as Facebook so nicely puts it. The information may be abused in aggressive predatory advertising, where people stuck between a rock and a hard place are targeted right where they hurt the most. Cathy O'Neil, Ph.D. in Mathematics at Harvard University, activist, and author of the book *Weapons of Math Destruction* (2016), points out that if someone is in possession of people's zip codes, demographics, habits, interests, and consumer preferences, they may use this information to effectively target ads specifically to people under social and economic pressure. If you have trouble making ends meet, you get fast and furious offers of payday loans at extremely high interest rates. If you are stuck in a steady job with little chance of climbing the career ladder, you are offered courses at expensive universities. The idea behind predatory advertising is:

> … to localize the most vulnerable persons and use their private information against them. This involves figuring out where they hurt the most, their so-called pain point.[10]

Vulnerable people are subjected to "false or overpriced promises"[11] by leveraging their weak points. It is documented that data brokers have sold lists consisting of possible "targets" for predatory advertising of snake oil or worse that include rape survivors, addresses of Domestic Violence Shelters, senior citizens suffering from dementia, HIV/AIDS sufferers, people with diseases and prescriptions taken (including cancer and mental illness), and people with addictive behaviors and alcohol, gambling, and drug addictions.[12]

---

[10] O'Neil (2016: pp. 72–73).
[11] Ibid. p. 70.
[12] Report to the General Assembly of the Data Broker Working Group issued pursuant to Act 66 of 2017, december 15, 2017, verified 29.06.2018: http://ago.vermont.gov/wp-content/uploads/2018/02/2017-12-15-Data-Broker-Working-Group-Report.pdf

Data-borne precision marketing is also exploited in political campaigns and ads. If you know the voters' profiles, it is much easier to persuade, seduce, or manipulate them and hence influence which candidate they vote for or if need be make them stay at home on election day. With the right data, you may be able to modify behavior and maybe even influence election results.

Money may buy you both the attention of voters and the information needed to influence their behavior in the desired direction. Barack Obama's campaign did it as early as 2008, when digital micromarketing became a big thing in American politics. Over a billion targeted e-mails were sent, particularly to young people and members of minorities in order to mobilize them to vote for the first time and vote for Obama.[13] Targeted political micromarketing reached a new level and took a dark turn in the Brexit referendum in the UK and in the 2016 Presidential Election in the USA. Both Leave.EU and Trump's campaign hired the firm Cambridge Analytica, which marketed itself as "using data to change audience behavior" in both commercial and political advertising.[14] When the *Facebook-Cambridge Analytica Data Scandal* broke recently, it came forth that Cambridge Analytica in 2014 started scraping personally identifiable information of up to 87 million Facebook users without their knowledge or consent.[15] The numbers are possibly even higher. With sufficient data about the electorate, it is possible better to manage it emotionally praying on pain points. This may be put to shady use as part of a "voter disengagement" tactic to demobilize the opponent's supposed supporters, so they do not

---

[13] Nisbet, M. (2012): "Obama 2012: The Most Micro-Targeted Campaign in History?", *Big Think* April 30, 2012. Verified June 24, 2017: http://bigthink.com/age-of-engagement/obama-2012-the-most-micro-targeted-campaign-in-history

[14] *Cambridge Analytica*, verified June 10, 2017: https://cambridgeanalytica.org/

[15] "The Cambridge Analytica Files," *The Guardian*, 2018. Verified, June 13, 2018: https://www.theguardian.com/news/series/cambridge-analytica-files

vote at all. This tactic is reported to have been employed in the American presidential election to discourage African-Americans to vote for Hillary Clinton.[16] Another tactic is to fuel anger and tensions, divisions, and conflict to the benefit of one's client. This method seems to have been employed in Kenya, where a lot of extremely divisive political messaging and targeted misinformation packages were observed during the 2017 election. Opening Pandora's box of political targeted micromarketing leveraging pain points may not only be damaging to the civility of democratic deliberation and participation. It may pose a danger to peace and stability. As Lucy Pardon, Privacy International Policy Officer, notes:

> The potential data-gathering could be extremely intrusive, including sensitive personal data such as a person's ethnicity. In a country like Kenya, where there is history of ethnic tensions resulting in political violence, campaigning based on data analytics and profiling is untested ground fraught with great risk.[17]

Many developing countries and emerging economies are at least as sensitive as the USA and UK to data misuse, misinformation, and fake news operations. At the same time, and at rapid pace, these new territories have caught the eye of the attention merchants and their entourage of big data analytics and demographic profiling to potentially hit the developing countries where it really hurts: on political self-determination.

There are dismal and even dystopian prospects in an attention and data economy where companies collect and appropriate personal information, commodify users into products,

[16] Burns, J. (2018). "Whistleblower: Bannon Sought To Suppress Black Voters With Cambridge Analytica," *Forbes*, May 19, 2018, verified 29.06.2018:   https://www.forbes.com/sites/janetwburns/2018/05/19/cambridge-analytica-whistleblower-bannon-sought-to-suppress-black-voters/#61a56d707a95

[17] Mirello, N., Gilbert, D., and Steers, J. (2018). "Kenyans Face a Fake News Epidemic," *VICE*, May 22, 2018. Verified June 13, 2018: https://news.vice.com/en_us/article/43bdpm/kenyans-face-a-fake-news-epidemic-they-want-to-know-just-how-much-cambridge-analytica-and-facebook-are-to-blame

and employ the gathered information against the very users to efficiently manipulate and influence behavior (see Chap. 7).

There is a lot of money in politics. The bulk of the campaign gold is spent on buying attention and influence on radio, TV, and the Internet. However, precious attention may come for free. The attention politicians are able to secure through exposure and time allotted to speaking, making headlines and set the agenda on the mass media's news coverage come without charge.

Chapter 2
The News Market

2.1   The King of the Agenda

At a conference in San Francisco in February 2016, CBS
Chairman and CEO Leslie Moonves conveyed the following
pertaining to the US Presidential Election and Donald
Trump's candidacy:

> It may not be good for America, but it's damn good for CBS …
> Man, who would have expected the ride we are all on now? The
> money keeps rolling in, and this is fun![1]

This comment from the CEO of one of the "Big Three" TV
networks in the US created quite a stir and received criticism
from Trump's opponent in the primaries, Republican Marco
Rubio, who ran an advertising campaign featuring this comment to prove the media actively backed Trump. Moonves
later defended his comment, claiming it was a joke misunderstood taken out of context.[2] Whether it was said in jest or not
is not to say, but it sounds like an honest disclosure judging

---

[1] Bond, P (2016): "Leslie Moonves on Donald Trump: "It May Not Be
Good for America, but It's Damn Good for CBS," *The Hollywood
Reporter*, February 29, 2016, verified May 20, 2017: http://www.holly-
woodreporter.com/news/leslie-moonves-donald-trump-may-871464
[2] Werpin, A. (2016). "CBS CEO Les Moonves clarifies Donald Trump
'good for CBS' comment," *Politico*, October 19, 2016, verified May 26,
2017:   http://www.politico.com/blogs/on-media/2016/10/cbs-ceo-les-
moonves-clarifies-donald-trump-good-for-cbs-comment-229996

© The Author(s) 2019
V. F. Hendricks, M. Vestergaard, *Reality Lost*,
https://doi.org/10.1007/978-3-030-00813-0_2

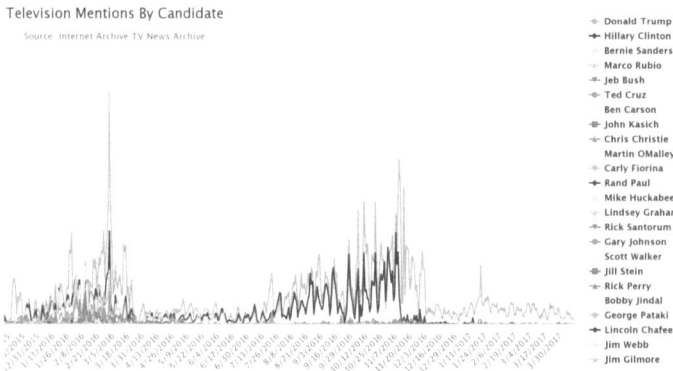

FIG. 2.1. Graph of the CBS television coverage of the US Presidential Election 2016 measured by the amount of times the candidate's name was mentioned. (*Presidential Campaign 2016: Candidate Television Tracker*, verified May 26, 2017: http://television.gdeltproject.org/cgi-bin/iatv_campaign2016/iatv_campaign2016?filter_candidate=&filter_network=AFFNET_CBS&filter_timespan=ALL&filter_displayas=RAW).

from CBS' coverage of the presidential election. Either way, the comment is spot on: CBS gave Trump much more coverage than his opponents throughout election season (Fig. 2.1).

And it wasn't just CBS. Donald Trump received the bulk of the news media's exposure and campaign coverage. He obtained a disproportionate amount of attention compared to the other candidates. As Trump himself said in a TV interview on the Fox News program *Your World with Neil Cavuto* on October 13, 2015, during the primary election cycle:

> I've spent zero on advertising because you and Fox and all of the others, I won't mention names, but every other network, I mean they cover me a lot, to put it mildly.[3]

---

[3] Donald Trump: "I've Spent Nothing on Ads Because of Fox News' and Other Networks' Constant Coverage," *Media Matters for America*, October 13, 2015, verified May 26, 2017: https://www.mediamatters.org/video/2015/10/13/donald-trump-ive-spent-nothing-on-ads-because-o/206115

Making it to the spotlight of mass media's news agenda provides precious media attention for free. Media attention is key to political success. Ever since the 1970s, a number of empirical media studies have confirmed the strong influence of the news media on public opinion. There is a clear tendency for people, events, and stories featuring prominently on the front cover or in TV and radio broadcasts to be perceived as the most important by the audience and, thus, the general public (McCombs and Shaw 1972; Dearing and Rogers 1996; McCombs and Reynolds 2002) (Fig. 2.2).

Without visibility on the news media scene, it is very difficult for politicians to influence the general public. News coverage and the visibility and exposure it leads to are vital resources for candidates in, or running for, political office. However, news coverage is also a *limited* resource. There is only so much paper in a newspaper, only so much time in a news broadcast or radio program. This naturally limits how much information and how many sources or candidates the media may devote its attention to. The people and stories that get news coverage are picked from a plenitude of potential options *at the cost* of the rest. The struggle to set the agenda of the mass media and get a share of this vital political resource is a zero-sum game where lots of attention given to one candidate means the others lose whatever the one gains.

Fɪɢ. 2.2. Information diet: the mass media's agenda becomes their audience's agenda.

Scorecard	Hillary Clinton Democratic	Donald Trump Republican	Gary Johnson Libertarian	"NOT TRUMP"	NONE OF THE ABOVE	DID NOT ENDORSE
Total Endorsements	**57**	**2**	**4**	**3**	**5**	**26**
Total Circulation	13,095,067	315,666	738,750	3,243,140	440,976	6,102,180
Endorsed Obama in 2012	40	0	1	0	0	3
Endorsed Romney in 2012	14	2	3	1	4	7
Endorsed Johnson in 2012	0	0	0	0	1	0
Did not endorse in 2012	2	0	0	2	0	15
SPLIT in 2012	1	0	0	0	0	0

FIG. 2.3. Overview of American newspapers' recommendations for the US Presidential Election in 2016. ("2016 General Election Editorial Endorsements by Major Newspapers," *The American Presidency Project*, verified May 26, 2017: http://www.presidency. ucsb.edu/data/2016_newspaper_endorsements.php).

Perhaps this was a conscious campaign strategy for Trump. As early as 2013 at a meeting in New York with a number of important Republicans who wanted him to run for governor, Donald Trump is supposed to have formulated a strategy not just for winning that office but the White House simpliciter:

> I'm going to suck all the oxygen out of the room. I know how to work the media so they never take the spotlights off me.[4]

Sucking all the oxygen out of the room is a well-known metaphor in American politics. It refers to attracting all the attention, leaving none of that vital reserve to others. The reality TV star Trump claimed that he could get the news media to dance to his tune and without charge score the attention other candidates had to pay great sums to get. He was right. But the media did not grant him their attention because they liked him and wanted him for President; almost all large, established mass media outlets were politically against Trump or turned on him during the campaign.

---

[4]Stokols, E. & Schreckinger, B. (2016). "How Trump Did It," *Politico*, February 1, 2016, verified May 26, 2017: http://www.politico.com/magazine/story/2016/02/how-donald-trump-did-it-213581

2.2   The Unwanted Candidate

The established news media did not support Trump for President. A lot of the airing was negative, and most established media recommended his opponent, Hillary Clinton. It's striking how massively the printed press supported Clinton over Trump (Fig. 2.3).

Even die-hard Republican newspapers such as *The Arizona Republic* and *The San Diego Union-Tribune* recommended a Democratic candidate for the first time ever; and *USA Today*—which had never before recommended any candidate—recommended *not* voting for Trump.[5]

The same negative view of Trump may be found all around the American TV landscape. Since CNN went on air in 1980, a variety of TV stations and news channels large and small have arisen on cable TV. The cable channels do not compete for the same broad audience as the original three large TV networks in the USA (CBS, NBC, and ABC). Rather, they direct themselves at different viewer segments and deliver programs catering to specific interests: sports, music videos, animal programs, sci-fi, and history. If there is an audience with special television needs, you can be sure there is a cable channel to cater to that need: The market will provide. The same goes for biased news coverage.

The cable news channel with the largest TV audience in 2016 was Fox News.[6] The station has obtained a monumental commercial success by producing and offering biased news coverage for a large audience of right-wing conservatives. It is not hard to argue and demonstrate that Fox News does not quite live up to its slogan of being "fair and balanced." An

---

[5] *USA TODAY*, Editorial, September 29, 2016, "USA TODAY's Editorial Board: Trump is 'unfit for the presidency'," verified May 26, 2017: https://www.usatoday.com/story/opinion/2016/09/29/dont-vote-for-donald-trump-editorial-board-editorials-debates/91295020/

[6] Schneider, M. (2016). "Most-Watched Television Networks: Ranking 2016's Winners and Losers," *Indiewire*, December 27, 2017, verified June 24, 2017: http://www.indiewire.com/2016/12/cnn-fox-news-msnbc-nbc-ratings-2016-winners-losers-1201762864/

example of this bias is the coverage of the party conventions in 2004, when the channel was completely uncritical of the Republican candidates. They procured more speech time and visibility than the Democrats who were routinely criticized by the hosts (Morris and Francia 2009).

News production for a specific audience so biased that it is possible to document may all the same be a very good business indeed. Others have taken note of that, and at the other end of the spectrum, left-wing channel MSNBC tries to imitate the concept. MSNBC likewise has an issue with balance and fairness at the other end of the ideological spectrum.

Cable TV has added to the polarization of the media supply and consumption. Political groups now watch different, biased news channels and programs. Nonetheless, Trump managed to fall out of favor with everyone. Even Fox News turned on him when he launched a misogynist attack on one of their hosts, Megyn Kelly, in which Trump was strongly insinuating that Kelly was driven by female hormones (to put it mildly). Ironically, his attack was a response to her critical questioning pertaining to his view of women.

No matter how critical TV channels ranging from Fox to MSNBC were of Trump's candidacy, this did not make them cover him any *less* up to the elections. He got the mass of attention all the same.[7] Whether they liked him or not, the *story* was simply too good to miss out on. Trump's spectacular campaign attracted too many viewers to turn off the cameras and the Trump talk. Even though the established mass news media were generally opposed to Trump's politics, he had stronger forces on his side. The market forces of commercial news and the resulting media logic.

---

[7] *Presidential Campaign 2016: Candidate Television Tracker*, verified May 26, 2017: http://television.gdeltproject.org/cgi-bin/iatv_campaign2016/ iatv_campaign2016?filter_candidate=&filter_network= AFFNET_CBS&filter_timespan=ALL&filter_displayas=RAW

2.3   Media Logic

When news media select, leave out, or produce news stories, it happens according to criteria defining "the good story." Those criteria are constitutive of media logic (Esser and Matthes 2013). Media logic is a set of rules and norms for behavior and action in the media: an institutionalized way of doing things, a modus operandi, or a set of guidelines for media professionals that consciously or unconsciously instruct them while selecting and producing the stories put on their agenda.

Media logic runs along three axes weighted differently by different media institutions and environments: journalistic ideals, commercial interests, and technological conditions.

1. Journalistic ideals

A news story should be true, meet the criteria of news value and anchored in journalism's role and self-perception as one of the central pillars in democracy, the free press. The media has a duty and a social obligation to inform citizens about important matters. The press must shed light on societal problems and adopt the role as gatekeeper of the public debate, guarding its quality by screening out lies, falsehoods, and nonsense installing checks and balances on the information finding its way to the public. Simultaneously the media has to act as watchdog of the powers that be, holding them accountable to the public while revealing possible abuses of power. The media's key role in democracy demands that journalistic values be upheld, such as independence, balanced reporting, transparency, integrity, truthfulness, and accuracy.

2. Commercial interests

From a commercial perspective, a news story is a good one if it attracts a large audience of readers, listeners, or viewers— if it attracts attention. To focus on the spectacular and dramatize the news may be an efficient way to achieve the commercial aim of a reaching a big audience. This makes the news coverage converge on entertainment. Joseph Pulitzer, who had a prestigious journalistic prize named after him, pointed out as far back as 1904 that there is a conflict between journalistic information ideals and commercial interests. He noted that they pull in opposite directions: one toward the responsibility to inform the public and the other toward the responsibility to create profit for the stockholders (Siebert et al. 1956). What works well in the market is not necessarily identical to what is good for democracy. A rephrasing of Leslie Moonves's previous comment up front amounts to: What is good for CBS stockholders is bad for American democracy.

3. Technological conditions

Media technology institutes the material framework for what works in specific media formats. A news story on TV requires good images. Political messages should preferably be expressed in short sound bites. The very type and format of a medium influence which messages reach the public and how they do it. Media theorist Marshall McLuhan puts it suc-

FIG. 2.4. The first-ever televised presidential debate: Nixon and Kennedy 1960.

cinctly in a famous petition: the medium *is* the message (McLuhan and Flore 1967). A case in point is the first TV-transmitted election debate in America, the debate between Kennedy and Nixon in 1960 (Fig. 2.4).

After the debate, a poll showed that the majority of radio listeners pointed to Nixon as the winner of the debate. TV viewers on the other hand named Kennedy the winner. This poll has since been criticized for resting on a statistical base to meager (Campell 2016). Nevertheless, later experiments have shown that radio listeners tend to value agreement with the candidate over political content when assessing debates, while TV viewers focus more on personality (Druckman 2003). Looks and charm in politics simply count for more on TV. Kennedy chimed in himself after winning the election:

It was TV more than anything else that turned things around.[8]

On commercial TV, personality, image, and fast retorts may dexterously replace political substance. All these aspects fit into the medium's image-based format and attract more viewers. News *and* politics run the risk of a reduction to pure entertainment.

2.4   Entertainment as Ideology

Neither political bias against Trump nor journalistic social responsibility outweighed the commercial motives prioritized in the media logic of the American news market. Whatever is the case for marketing also goes for news production: When the point is predominantly to reap attention, entertainment is a most efficient method to catch it and keep it. The entertainment value of a story or a program often enough beats all other criteria. Entertainment attracts a hefty audience. In a purely commercial media market, it trumps the news media's and journalists' political views, biases, and ideological

---

[8]Webley, K. (2010). "How the Nixon-Kennedy Debate Changed the World," *Time*, October 23, 2010, verified May 26, 2017:http://content.time.com/time/nation/article/0,8599,2021078,00.html

positions. Media theorist and critic Neil Postman puts it like this in his classic book, *Amusing Ourselves to Death*:

> Entertainment is the overlying ideology of all discourses on TV.[9]

Ideology works in a structural manner and operates largely independent of a person's conscious decisions. A lot depend on habits and routines. Ad that, if CBS did not spotlight Trump, then other TV stations that did would make off with the viewers and, hence, the revenues from advertising. In a commercial media environment where news is already driven by its entertainment value, coverage and visibility may be obtained by providing just that: entertainment. Being sufficiently scandalous, rude, and politically extreme may keep all eyeballs on you with no attention left for your opponents. Trump succeeded in keeping the cameras pointing in his direction, and the audience enthralled. He tended to be spectacular, polemic, and dramatic, a recipe for good television.

Not only in the traditional media did Trump have the media logic on his side. It was so the case even more online and on social media.

2.5   The Free Online Market for News

The Internet has opened up a new market for news and broken the near monopoly of the mass media as producers and distributors of news and information. It is a consequence of the development of both information technology and digitization together with the decentralized information infrastructure the Internet offers. The digitization of text, audio, video, and graphics has changed the production conditions for media and news products: what used to be the former media *consumers* are turned into potential content *producers*. Everyone with a smartphone and appropriate software for the treatment or manipulation of images, sounds, and video may create media products that used to require a large and costly production apparatus the likes of which only the mass

---

[9] Postman (1985: p. 25).

media could afford. The Internet and social media have provided an infrastructure that makes it easy and cheap to publish and spread a message or a news story. This has turned citizens into potential civic journalists and made the genesis of online news, debate, and special interest platforms possible. The decentralized network structure has the consequence that users do not need to pass any journalistic gatekeepers in order to publish and distribute material. With work, effort and somewhat savvy of the tools of the digital trade, users may spread material virally and even anonymously if platforms are used permitting anonymity or else users may create fake profiles on platforms that do not.

In spite of the new conditions for news production, presentation and proliferation that have expanded the news market, the news *diversity* does not seem to have changed all that much. A study on the Twitter agenda (Bryan et al. 2014) concludes that there is a limited sum of stories and news that gets all the attention on the online news scene, just like in the mass media. Traditional news media, journalists, Internet media, and Twitter users copy or share each other's stories and news; the same few news stories are circulated again and again with small changes, alterations, and comments. Another empirical study of local Twitter environments in six different countries (Humprecht and Esser 2017) concludes that the stronger a country's public service media is, the more diversity the online agenda becomes. A completely liberalized and commercial news market does not necessarily display a greater diversity of news stories. The same goes for the diversity of the news sources that attract attention and get the ear. These conditions exist because online attention does not follow a normal distribution, but rather a power law distribution: A few players get the bulk of the attention (Hindman 2009; Webster 2014), while everyone else has to fight over the very limited attention at the tail of distribution. Much like the world economy: 1% has 50% of the world's wealth, and the other 99% have to fight over the remaining 50% (Fig. 2.5).

New alternative voices may indeed enrich journalism and public debate, but only if the new media respect and operate according to journalistic ideals and virtues. If not, they rather contribute to further division, polarization, and circulation of

The Power Law of Online Attention

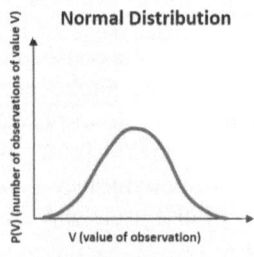
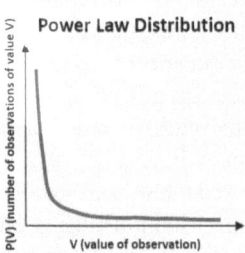

FIG. 2.5. The winner takes all, or, at the least, a very small number of players get the bulk of the attention on the Internet. A power law reins the attention distribution rather than a normal distribution.

misinformation. This was the case in the USA leading up to the 2016 election. The right-wing news platform Breitbart that became the voice of the extreme alternative right hit the jackpot of attention but disregarded journalistic ideals completely and continues today (Fig. 2.6).

The alt-right movement and platforms such as Breitbart managed to dominate the bulk of the online agenda. The stories circulated and were harvesting most attention on social media and were mainly in Trump's favor. A common denominator of the stories winning the battle for online attention is the lack of journalistic criteria for truthfulness and documentation:

"Pope Francis shocks world, endorses Donald Trump for president"
"Donald Trump sent his own plane to transport 200 stranded marines"
#Pizzagate
"Ireland is now officially accepting Trump refugees from America"
"WikiLeaks confirms Hillary sold weapons to ISIS … Then drops another bombshell"[10]

---

[10] Ritchie, H. (2016). "Read all about it: The biggest fake news stories of 2016," *CNBC*, December 30, 2016, verified April 5, 2017: http://www.cnbc.com/2016/12/30/read-all-about-it-the-biggest-fake-news-stories-of-2016.html

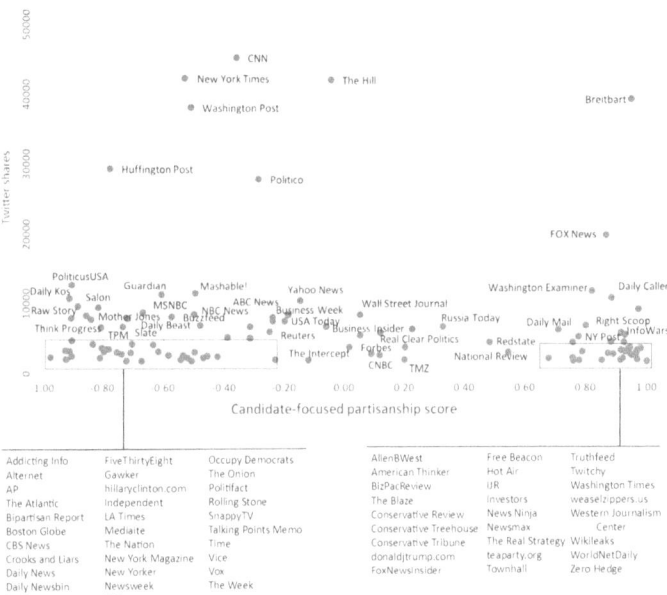

FIG. 2.6. The sources for the news stories that Clinton and Trump followers, respectively, shared on Twitter leading up to the 2016 election. (Benckler, Y., Faris, R. Roberts, H. & Zuckerman, E. (2017). "Study: Breitbart-led right-wing media ecosystem altered broader media agenda," *Columbia Journalism Review*, March 3, 2017, verified May 26, 2017: https://www.cjr.org/analysis/breitbart-media-trump-harvard-study.php).

Such misleading stories are not informing anybody of anything. Neither do they contribute to enrich the democratic deliberation. They undermine it.

Social media are better at spreading outrage than sound information and documented news stories. This is partly a result of the media logic of social media. Meeting the journalistic ideals is not an intrinsic part of the fabric online and on social media. The *ideal* is rather sharing of personal information with minute-to-minute posts and updates on whereabouts, opinions, thoughts, and feelings (Klinger and Svensson

2015). It is about connecting with others while expressing oneself. The *commercial imperative* for social network platforms aligns with this ideal of sharing and self-expression. When engaging and interacting online—in contrast to the passive consumption of mass media—both attention and user data are delivered to the companies to monetize and turn into profits. The digital *technology* makes all this possible by providing an informational infrastructure and designing for engagement, interaction, updates, and rapid peer-to-peer communication. In this environment, expressing and sharing one's outrage is king: To secure virality, create engagement driven by outrage via loud, spectacular, and angering information. This is not only a viable tactic for angry citizens or alternative extremist media outlets. Political leaders and candidates may also employ the strategy of outrage effectively. With social media, not only the citizens but also the elite has a new tool of communication that may be used to set the agenda and divert attention from criticism.

2.6   A Channel for the Elite

Even though the Internet and social media's open news market free of much gatekeeping have afforded a bullhorn to new voices and granted some of the politically incorrect and extremists a spot in the public debate, media research does not indicate that social media has generally shifted the power game between the political elite and the citizens in favor of the latter.

Facebook and Twitter have made it possible for citizens to debate politics independently of mass media and have created new possibilities for organizing and engaging online in democratic debate. This is not the same as actually having a say and confronting the powers that be or exerting much, if any, influence over the political agenda. Not only citizens may now surpass the journalistic gatekeepers; so, too, may the politicians. The social media that made it possible to speak directly to the politicians simultaneously gave the politicians a direct channel to their supporters and the general public.

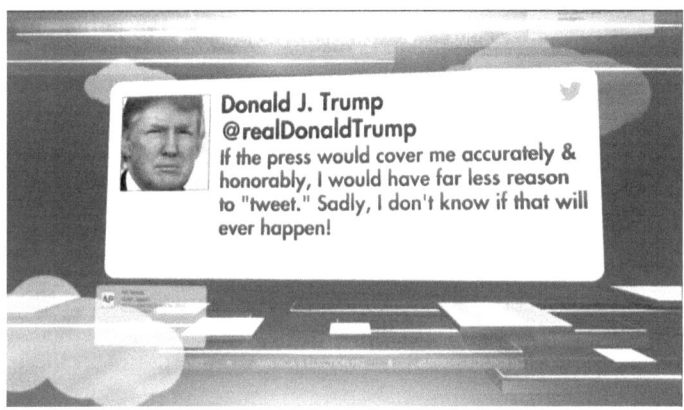

Fɪɢ. 2.7. The news story on Fox News about a tweet from Trump in which he cites the media's purported dishonesty and lack of accuracy as the reason why he tweets so much. (Should Trump Keep Tweeting to Counter 'Biased' Media Coverage?", *Fox News Insider*, December 6, 2016, verified June 14, 2017: http://insider.foxnews.com/2016/12/06/writer-says-donald-trump-tweets-because-media-covers-him-unfairly).

Therefore, social media may also act in favor of the powers that be. An empirical study on the use of Twitter as a news source in four Dutch and four British newspapers from 2007 to 2011 concluded the following:

> Elite sources may obtain more control with the public debate due to Twitter. News are no longer the product of a negotiation, but merely a result of one-way communication. Not being available to journalists […] in the middle of a media storm, but simply throwing them a tweet instead, as the Dutch right-wing politician Geert Wilders does, is an efficient strategy to controlling and framing the news discourse.[11]

Social media has become a news source for journalists. The resulting news coverage is often based on the tweets from the political elite (Skogerbø et al. 2016) (Fig. 2.7).

---

[11] Broersma and Graham (2013: p. 463).

This creates a short-circuited media where you may tweet (in a biased manner) about news you watch on television, and then your tweet is taken up by the mass media publishing it as a news story. With a Twitter account, you may troll the media and the public and suck all the oxygen out of the room. Trump exemplifies being a master of that craft, and he knows it.

Without tweets, I wouldn't be [in the White House].[12]

This situation puts the American news media in a dilemma. Since Trump is President, his tweets are almost per definition newsworthy. This makes his Twitter account an efficient instrument in the attention economy. With a limited agenda and attention, extreme and spectacular tweets may serve to distract the news media and the general public. The tweets take up both the news media's and the general public's limited attention that could otherwise have been spent on stories with more substance. It is often worth asking what is *not* receiving attention, while everybody stares at the spectacle.

[12] Barber, L., Sevasopulo, D. & Tett, G. (2017): "Donald Trump: Without Twitter I would not be here — FT Interview," *Financial Times*, April 2, 2017, verified May 26, 2017: https://www.ft.com/content/943e322a-178a-11e7-9c35-0dd2cb31823a

Chapter 3
Attention Speculation and Political Bubbles

3.1   The Mediatized Society

We live in an increasingly mediatized world. Mediatization refers to the tendency of societal institutions to be more and more dependent on the media and adapt themselves to its conventions and to media logic (Hjarvad 2008). In a mediatized society, the media veritably establish the conditions for social interactions and relationships, commerce and marketing, science and debate, and activism and politics. When political activists protest or organize a demonstration in order to send a political message, it is essential to get media coverage. The message must be heard by people and parties other than the activists themselves. There is no point in "Occupying Wall Street" unless documented and disseminated. Mediatization provides social and political actors with a strong incentive to act according to the media's precepts.

  In the mediatized society, it is key to deliver material to the media for *the good story* which will hit the agenda right way. Politics is mediatized when political actors such as cabinet executives, ministers, members, spin doctors, and press officers tailor their communication—and sometimes even their politics and legislation—to the news media's criteria for the good story. If the news media align their criteria to the commercial interests, the result may become, at its most garish, that entertainment value becomes the real ideal of news coverage.

© The Author(s) 2019   35
V. F. Hendricks, M. Vestergaard, *Reality Lost*,
https://doi.org/10.1007/978-3-030-00813-0_3

Mediatization has been seen as a symptom of the increasing power of the media. Not only do they set the agenda; their criteria for doing so spread to other institutions and actors as incentives to subjugate to their media logic. Mediatized politics has been characterized as politics that has lost its autonomy and independence relating to the media (Mazzoleni and Schulz 1999: p. 250).

The media logic may also be used to take back power from the media when the politicians adapt to it and use its structure and dynamics to their own advantage (Thesen 2013). This is what political spin is all about. The ploy does not necessarily include gaining mastery of the scene on a commercial media market, selling entertaining, riotous, and spectacular stories. It may also play on journalism's professional virtues like the noble invocation to report from the inner circles of power and let the public in on the end game going on behind closed doors. Planned tactical "leaks" may, for instance, be employed to play the journalists in this regard and set the agenda to one's benefit.

Alas, the consequence of politicians spinning the media too efficiently to their own end may be the emergence of a media-created political reality detached from the real world and its problems.

3.2   The Media-Created Political Reality

The media may not just cover the political reality; they may partake in creating it. When that happens, it may be seen as a symptom of what the late French philosopher Jean Baudrillard called *hyperreality*. Here, the divide between media and reality implodes, and what goes for "reality" becomes a media product or a *simulation* of reality which is impossible to discern from the real thing (Baudrillard 1994). Fiction may acquire such a marked influence on reality that the distinction between facts and fiction, between news and entertainment, gets blurred. Or, as the Danish satirical Nihilistic People's Party described it back in 2010:

> The entire political reality is a media-created illusion the politi-
> cians maintain in order to hide the fact that the project of
> Enlightenment ended in Dancing With the Stars.[1]

This way of putting it does not seem overly absurd anymore,
nor does Baudrillard's philosophy seem that radical. One
case in point is how an episode of the world famous Danish
TV series *Borgen* thematizing prostitution was used as political leverage for a bill to decriminalize prostitution in the *real*
Borgen, the Danish Parliament at Christiansborg. Or worse:
Trump's fabled way into American politics.

3.3 Where Was Obama Born?

Donald Trump admitted at long last on September 16, 2016,
that former US President Barack Obama was rightfully an
American citizen. Prior to that date, Trump had for years
advanced an argument, later turned into an extensive conspiracy theory, to the effect that Obama not being born in the
USA, not being an American citizen for real, couldn't legiti-
mately hold the Presidency. It was virtually this crazed claim
and debacle that marked Trump's entrance into American
politics.

What fostered these speculations was the fact that Obama's
father was an economist in Kenya, which was taken to show
that Kenya, and not Hawaii, had to be Obama's true birth-
place. Another story had it that Obama at one point acquired
Indonesian citizenship and lost his American citizenship in
the process. And then there was the story of how dodgy it
surely seems that Barack Obama's middle name is Hussein.

For years Trump sat at the conspirational hearth and
refrained from shutting down the rumor about Obama's foreign citizenship. He actually boasted that it was he who
forced Obama to prove in 2011 to be rightfully American by

---

[1]Vass, T. & Vestergaard, M. (2010): Nihilistisk Folkeparti [Nihilistic
People's Party], front cover. Verified June 24, 2017: http://www.nihilis-
tisk-folkeparti.dk

An 'extremely credible source' has called my office and told me that @BarackObama's birth certificate is a fraud.

↩ Reply   ⟲ Retweet   ★ Favorite

690
RETWEETS   **115**
FAVORITES

4:23 PM - 6 Aug 12 - Embed this Tweet

FIG. 3.1. A tweet from Donald Trump in 2012 after Obama had presented his birth certificate, and the "birther" case against Obama continued.

showing his birth certificate. That, in turn, had some people doubt the veracity of the document, and a new conspiracy theory was born (Fig. 3.1).

But then something happened: While Trump refused to acknowledge the sitting President Obama's American nationality, the Trump campaign may have started to realize that the birther case was not a strong case. Trump then admitted that President Obama was indeed an American citizen but at the same time claimed that the rumor originally was set to sea by Hillary Clinton's presidential campaign back in 2008 when Clinton and Obama competed internally to win the Democrats over. Trump also managed to frame his backing down in such a manner that it sounded as if he was doing Americans and Obama himself a huge *favor* by finally dropping the birther business.

Over the years, this case has enjoyed a lot of attention, even though it turned out to be false. Trump managed for a time to benefit from the fact that entertainment value is a decisive news criterion in a commercialized media market. On The Daily Show in 2013, John Oliver called upon Trump to run for President for the fun of it. Oliver has since lived to regret it.[2]

Not all politics is like that. All the same, politics is not entirely above the delicate and quite dangerous problem the mediatization has created: Politics may move further and further away from reality and facts with all their inherent complexities. The citizens lose out in the long run if the negative consequences of policies and legislation are ignored politically or if attempts are made to hide them from the public, other politicians, and the media.

If politics is all about winning the horse race and voter maximization, reality may lose out. One would think that politics is supposed to solve real problems. It should be considered a victory to ensure, say, that less environmental damage is caused, independently of whether more or less support for the political party comes along with this triumph. If there is no ambition to solve real problems, then politics is reduced to showbiz and speculation in sending, creating, or struggling over political signals and symbols.

3.4 Signal Legislation and Symbol Politics

The context is different, but "signal legislation" are bills:

1. Whose primary goal is to signal a specific point of view
2. Proposed with no real interest in the consequences
3. Often, with no interest in the real scope of the problem (Elholm 2011)

Signal legislation has the potential for political speculation as it is immune to, and detached from, what other experts or science might have to say. The point is not about the facts, the particulars, or calculating the consequences of a legislation

---

[2]THR Staff (2016): "John Oliver Regrets Begging Donald Trump to Run for President," *The Hollywood Reporter*, November 7, 2016. Verified June 10, 2017: http://www.hollywoodreporter.com/news/john-oliver-donald-trump-president-944682

but simply the signal it sends. You signal your point of view. When fact-finding is not a concern, the legislative process may be carried out that much faster without commissions, committees, or other forms of investigative bodies: Fast process, fast signaling of political determination, trustworthiness, initiative, and dynamism. And, as Elholm points out:

> The success criterion for signal policies is pure genius. The results of the legislation are secondary. The purpose of the law has been reached as soon as the signals are received, or perhaps even when they have been sent![3]

No matter how comfortable signal policies and legislation may seem, they come at a cost. The lack of calculating the consequences may mean a whole array of measures are taken that are futile at best and may have unintended social, economic, or administrative consequences. Arguing for signal policies often requires a grasp of the majority's sense of justice, opinions, set of values, or sentiments, and those are hardly as clear-cut as the rhetorics would often have it. Additionally, such signals may cause a polarization between those whose opinions the signals suit and those who are stigmatized or outraged by the very same signal. Finally, focus and resources are taken away from more substantial efforts to come up with concrete, effectual solutions to societal issues through legislation and policies. Signal legislation is the fast, easy, or tactical political reaction that does not necessarily lead to the durable, efficacious, or sustainable solution.

A case in point is the veiling ban passed in Denmark. Generally speaking, the ban applies to all forms of covering your face in public space. In other words, the ban does not specifically address religious veiling. Nevertheless, the Danish foreign minister was less diplomatic and more blatant in a statement that he posted on his Facebook profile. In the statement, he makes it clear that the ban is aimed directly at niqab and burka, combating the "dark men" that are responsible for the oppression of women symbolized by garments

---

[3] Elholm (2011: p. 170).

that cover their faces.[4] But already before the ban, Danish legislation allowed for sentencing an individual to prison for up to 4 years, if that somebody is forcing another to cover his or her face.[5]

Signal legislation is often carried out without much consideration for the actual scope of the problem the law is addressing. Thus, the last investigation of the use of burka and niqab in Denmark executed in 2009 reveals that as few as 100–200 individuals are actually wearing such garments. About 50% of them have converted to Islam.[6]

At the same time, Denmark is not showing any interest in the effect of a similar law implemented in France in 2011. According to the findings of sociologist Agnés de Féo, the French veiling ban has been counterproductive rather than beneficial to integration. The women who nowadays are fully covering their face started doing so after the implementation of the law. And women who fully covered their face prior to the implementation of the veiling ban are no longer leaving home.[7] Those kinds of investigations and evaluations of actual impacts are of no concern to the politician who is pressing for legislation of purely symbolic value. Jonathan Laurence, Professor of Political Science at Boston College and expert in Muslim communities in Europe, says it tellingly:

> I don't think policymakers would pay such studies any mind since these laws are never about integration effects.[8]

---

[4] Samuelsen, A. (2017). *Facebook*, October 6, 2017 (Danish), verified December 2, 2017: https://www.facebook.com/AndersSamuelsenLA/posts/10156454217557366

[5] https://www.retsinformation.dk/eli/ft/200912L00181

[6] "Rapport om brugen af niqap og burka"(2009). Verifiziert 02.12.2017: https://www.e-pages.dk/ku/322/html5/

[7] Taylor, A. (2016). "Banning burqas isn't a sensible response to terrorism," *Washington Post*, 12.08.2016, verified 26.06.2018: https://www.washingtonpost.com/news/worldviews/wp/2016/08/12/banning-burqas-isnt-a-sensible-response-to-terrorism/?utm_term=.8283b0b0bde5

[8] Ibid.

Denmark wants to seize jewelry and cash from refugees:

Denmark wants to seize jewelry and cash from refugees
The law "has been branded petty and cruel."
washingtonpost.com

11:35 AM - 18 Dec 2015

23 Retweets **8** Likes

♡ 4   ⟲ 23   ♡ 8

Fig. 3.2. The Danish "jewelry law" went worldwide and made headlines in international media like *The Washington Post.*

Signal legislation is a close cousin to politics of symbolism, which also is more about showing vigor than actually solving problems. A case in point is Danish legislation dubbed the "jewelry law" that was enacted in early 2016 (Fig. 3.2).

An entire immigration law package was passed, but it was the call for the confiscation of refugees' possessions that got the bulk of the attention, both nationally and internationally. The attention was of negative nature. It fueled outrage.

When push came to shove, the warrant to confiscate possessions was only utilized four times during the first year the law was in force. However, other parts of the law package that did not get much attention had significant consequences. L87, of which the "jewelry law" forms a part, entails that certain refugee groups have to wait up to 3 years before they may even file for a family reunification, let alone have their loved ones *actually* joining them. The time limits established by international conventions are bent to their extreme. Be that as it may, not a lot of fuss was made pertaining to this part when the law package was passed and later enacted in its entirety.[9]

If political signals and symbolic messages are often most proposed with scarce interest for their effects and the real scope or nature of the societal issue, then why send them at all? Because they serve an entirely different purpose. Years ago, the influential political scientist Murray Edelman put it this way: They may condense a whole political, ideological, or cultural narrative in a very simple manifestation, because:

> Condensation symbols evoke emotions bound to the situation. In a symbolic incident or action, they condense patriotic pride, worries, memories of yesteryear's victories or humiliations, promises of future grandeur; some of these or all of these simultaneously.[10]

Signal legislation and symbol politics are ways to speculate in the attention market to create political bubbles.

3.5   Political Bubbles

Speculative bubbles emerge now and again on financial markets. Fairly recent examples are the housing bubble bursting in 2008 causing the financial crisis and the dot.com bubble around the arrival of the millennium. History offers a list of financial bubbles, dating all the way back to the first documented, speculative bubble: the Dutch tulip bulb bubble in

---

[9] Olsen, T.L. (2017). "Et år med omstridt smykkelov: Politiet har brugt den fire gange" *DR.dk* January 16, 2017. Verified May 9, 2017: https://www.dr.dk/nyheder/politik/et-aar-med-omstridt-smykkelov-politiet-har-brugt-den-fire-gange
[10] Edelman (1979: p. 847).

1636–1637 (Brunnermeier and Schnabel 2017). When financial bubbles emerge, they are often driven by stories of huge monetary gains waiting ahead, accompanied with narratives to the effect that *this time around*, it will be different than the previous bubbles after which investors and speculators suffered great losses when the bust came (Schiller 2017). This kind of narrative may trump more realistic assessments of the worth of a given asset. If the narratives are not anchored in economic reality and realistic expectations, then the forming of the bubble may be described as a "collective rejection of reality" (Quigging 2010: p. 132). Bubbles on financial markets are defined as situations in which financial assets are systematically traded at prices far exceeding their fundamental value (Vogel 2010). An asset's fundamental value is the dividend expected in the long run if you were to keep it. If tech-shares, toxic mortgages and subprimes, or tulip bulbs are traded at prices above and beyond a realistic assessment of their worth, then the price has inflated artificially through speculation and no longer represents the underlying worth of whatever the asset being traded. If unaware of the fundamental value of an asset, investors might just end up paying far too high a price for very little or for nothing at all.

The same thing may happen in attention economy of the political sphere. The emergence of political bubbles on the attention market may also be described as a collective loss of reality. The political substance disappears ever more from the equation, just like an assets' fundamental value may do in financial bubbles. On the attention market, circulating stories are not just one of the reasons for the formation of bubbles. The stories *themselves*, or the events that reach the news agenda, are what may become bubbles. The characterization of financial bubbles may be transferred to attention economics by replacing:

- *Asset* with *news item*
- *Price* with *amount of attention*
- *The fundamental value* with *political substance*

Thus, a political bubble is defined as a situation in which a political item gets a measure of attention in the media far exceeding what the political substance justifies.

3.6   The Substance

Citizens and journalists alike, as well as politicians (especially if they are under pressure), often call for political substance. Political substance concerns solving societal problems and improving social, economic, or cultural conditions in the long run. Politics is much about creating results; otherwise, it simply crumples to reality TV, horse races, and drama. The fundamental value of a political item—its political substance—may be established by its capacity to *represent* a societal problem. By being placed on the agenda, light may be shed upon it; it may be deliberated and acted upon. Trivially there are ideologically founded differences among various political parties pertaining to what should be considered societal problems and their order of priority. In Denmark, for instance, equality is seen as more of a problem by the left wing than by the right wing, and the opposite goes for high taxes. However, there is still quite a consensus that political issues such as unemployment, immigration, crime, and domestic violence *are* indeed social problems that need solving. These problems will not get solved politically, if the political debate loses its substance. If too much attention is spent on symbolic messages, signal policies, and debating for the sake of debacle without any potential of reaching concrete results, other than perhaps a further polarization of the political fronts. Swaths of the scarce attention in such cases go to political items with no actual content.

What is being debated and sucks all oxygen out of the public space in a political bubble is not, for instance, how best to solve integration problems or address the current refugee and immigration crisis. Rather, it is the sending of a signal to the public germane of *who* the sender is and where that person or party stands politically, if there were to be some who didn't know this already.

Political bubbles thrive particularly well in polarized opinion environments where the good story or the recognizable symbol gives easy and direct access to being *for* or *against* something: for or against the politics of Washington, for or against the EU, etc.

In these polarized opinion environments, you may also *angle for a scandal* and cash in on the opponents' anger. The angrier the other wing, the more sturdy and perhaps even vigorous you may seem to your followers. Thus, politicians may obtain an advantage from stimulating bubbles with spectacular, controversial, and provocative messages. Political bubbles may grant speculatively yield on the image and identity front, but not have much yield to society in terms of tangible political results, rather quite the contrary. The latter would require real political *investment* in substance, not speculating in the market of attention for political products.

3.7   Investment and Speculation

The British economist John Maynard Keynes (1883–1946) defined the difference between speculation and investment:

> Investment is an activity that predicts an asset's return during its lifetime, whereas speculation is an activity that predicts the market's psychology.[11]

Financial investment involves assessing the fundamental value of assets, while speculation may ignore the value and play on the market. The return on an investment, however, also depends on various market fluctuations. Hence the dissociation between investment in the long run and speculation in the short run is not absolute. But that the difference

---

[11] Quoted from Peterson (2016).

between investment and speculation is not absolute does not mean it is not there at all. *Political* investment embroils existing knowledge and information about a policy to try to assess its expected returns for society while determining the effects and side effects if the policies were to be carried out. If such assessment is not exercised, action is taken blindfolded. The fact that the future cannot be predicted exactly and in detail does not prevent the tooling of useful scenarios based on the current accessible evidence. Take climate models, for example:

> These models are designed to provide scenarios of the future that can be used for guiding decisions about what policies to follow, such as how to reduce undesirable climate impacts and build resilience, and to plan for the future in the most cost-effective ways. Because human actions themselves are not predictable, these are not predictions, but rather they are called projections and depend on the nature of the "what-if" question.[12]

This argument goes for all the global challenges that the world faces, from climate and migration to online misinformation. Without acknowledging and using evidence, there will be no real solutions. If politics is to be more than show, speculating in attention and scandal and political entertainment with or without spin, it *has* to be informed by available evidence.

The same is true for the actual difficulties regarding receiving seekers of asylum, immigration, and integration of folk and fugitives. When professionalism, experience, and scientific evidence are replaced with symbolic politics coupled with the value of signals and speculation in attention, the fight against parallel societies, ghettoization, social control, radicalization, extremism, and terrorism hold only marginal chances of successful solutions.

---

[12]Trenberth, K. & Knutti, R. (2017): "Yes, we can do 'sound' climate science even though it's projecting the future," *The Conversation*, April 5, 2017. Verified May 10, 2017: http://theconversation.com/yes-we-can-do-sound-climate-science-even-though-its-projecting-the-future-75763

So how are we doing? Well, it sure is uphill. Evidence from anthropogenic global warming is not the only kind of evidence being called into question on political grounds; the weather itself gets contested. Not too long ago, it was turned into a political question whether the sun was shining or not on a special occasion.

Chapter 4
Alternative Facts, Misinformation, and Fake News

4.1 Alternative Facts

Some very specific topics stole the show the day Trump was inaugurated as president on January 20, 2017. The limelight topics are related to simple questions about the facts of the day: Did the sun shine or not during the inauguration speech? How large was the crowd? Was this crowd larger or smaller than the one present at President Obama's first inauguration? It seemed clear from available photos and video footage that the sun did not shine at any point during Trump's speech. Nonetheless, Trump claimed otherwise later that same day during his speech at Langley Air Force Base:

> The rain should have scared them away. But God looked down and he said, "We're not going to let it rain on your speech." In fact, when I first started I said, "Oh no." First line, I got hit by a couple of drops. And I said, "Oh, this is, this is too bad, but we'll go right through it." But the truth is that it stopped immediately. It was amazing. And then it became really sunny, and then I walked off and it poured right after I left.[1]

Trump's first act in office was to make a claim which could easily be proven to be false. All it took was to look at the

---

[1] Sharman, J. (2017): "Donald Trump: All the false claims the 45th President has made since his inauguration," *The Independent*, January 23, 2017. Verified February 4, 2017: goo.gl/0TyieB

© The Author(s) 2019
V. F. Hendricks, M. Vestergaard, *Reality Lost*,
https://doi.org/10.1007/978-3-030-00813-0_4

images from the event to see whether the sun shone or not during his speech.

Another factual question was the size of the crowd. Topping up his speech at Langley, Trump made the crowd size a moot issue:

> We had a massive field of people. You saw that. Packed. I get up this morning. I turn on one of the networks and they show an empty field. I say, "Wait a minute. I made a speech. I looked out. The field was … It looked like a million, a million and a half people." Whatever it was, it was. But it went all the way back to the Washington Monument.[2]

Beyond much doubt, photo material from the inauguration reveals that the crowd did not extend that far. Numbers published by the transportation authorities in Washington, D.C., showed 570.557 registered travels during Trump's inauguration, while 1.1 million travels took place around Obama's first inauguration and 782.000 related to his second inauguration (Fig. 4.1).

This did not keep former White House Press Secretary Sean Spicer from launching a frontal attack on the press accusing it of "reporting erroneously on purpose" with reference to size of the crowd. On January 21, 2017, Spicer stated: "This was the largest audience ever to witness an inauguration, period, both in person and around the globe."[3] Later that same day, Kellyanne Conway chimed in. On NCB's *Meet the Press*, this key presidential advisor was confronted with publicly accessible testimonies contradicting Spicer's statement. She defended the statement by arguing it was neither a lie nor a falsehood; rather, Spicer was conveying "alternative facts." On January 23, 2017, Spicer said something to the same effect at a press conference:

> Sometimes we [The White House] can disagree with the facts.[4]

[2] Ibid.

[3] Ford, M. (2017): "Trump's Press Secretary Falsely Claims" *The Atlantic,* January 21, 2017. Verified February 4, 2017: goo.gl/HvnFUY.

[4] Smith, D. (2017): "Sean Spicer defends inauguration claim." *The Guardian,* January 23, 2017. Verified February 4, 2017: goo.gl/djwg7x.

F<small>IG</small>. 4.1. The human crowd gathered at Obama's inauguration in 2009 to the left compared to the one during Trump's inauguration in 2017 on the right.

Leave that one on the screen for a minute. Is it even feasible to deny facts that may be readily verified, propose "alternative facts," or "disagree with the facts?" Logically, the term "alternative facts" makes no sense at all. It breaks the Aristotelian principle of noncontradiction, which probably has been in vogue as long as we have been able to think. As journalist Chuck Todd from NBC made clear to Conway, it is not possible to propose "alternative facts" without either being mistaken or downright lying. Truth be told; Conway did indeed use the term "alternative facts." Apparently that was not what she meant to say. She meant to say Spicer presented "alternative information" as to the number witnessing the inauguration: 2 + 2 = 4 but so does 3 + 1.[5] Conway was never

[5] Fitzgerald, S. (2017). "Conway: Spicer was using Alternative Information," Newsmax, 24.01.2017, verified 15.06.2018: https://www.newsmax.com/politics/conway-spicer-alternative-information/2017/01/24/id/770122/

Fig. 4.2. The interview in which Kellyanne Conway introduces "alternative facts".

really able to get that message across. The "alternative facts" stuck in the media perhaps because it seemed indicative of so much else of what was going on.[6]

Be that as it may, facts are facts. Statements regarding factual matters are either true or false. An alternative claim denying a true statement is simply a false one. You may of course disagree that these *are* the facts; but disagreeing *with* facts is to disagree with reality. The statement "the sun is shining" is only true if the sun is actually shining, regardless of specific preference political or otherwise (Fig. 4.2).

Some factual issues are easier to settle than others. Sometimes the facts and truth are difficult to establish. Sometimes they have not been found yet or seem elusive. That's why science and journalism alike keep inquiring.

---

[6] Pengelly, M. (2017). "Kellyanne Conway: 'alternative facts' was my Oscars La La Land blunder," The Guardian, 04.03.2017, verified 15.06.2018:   https://www.theguardian.com/us-news/2017/mar/03/kellyanne-conway-alternative-facts-mistake-oscars

4.2   Truth in Science and Journalism

The goal of science is to research, understand, and explain the world and the way it works as far as the existing scientific methods will allow. Science seeks truths and knowledge about nature, social conditions, humanity, and technology.

The self-declared goal of science does not fall far from that of journalism. According to the principles stated in the *Project for Excellence in Journalism*, the primary obligation of journalism is to the truth, but not in an absolute sense. This journalistic truth "is a process that begins with the professional discipline of assembling and verifying facts. Then journalists try to convey a fair and reliable account of their meaning, valid for now, subject to further investigation. Journalists should be as transparent as possible about sources and methods so audiences can make their own assessment of the information."[7]

The world is constantly evolving, and new information may come forth, and thus journalistic truth, just like scientific truth, is tentative. Journalism has to be as transparent as possible when it comes to sources and methods making it possible for readers to make up their own minds about their views and the correctness of the information put in front of them.

Journalism should also be careful to avoid bubble formation and always keep news in proportion but not leave out important details. *Journalist's Resource* notes that "Journalism is a form of cartography: it creates a map for citizens to navigate society. Inflating events for sensation, neglecting others, stereotyping or being disproportionately negative all make a less reliable map."[8] Sensationalism and poor journalism, as well as pseudoscience, put the reliability of the map under pressure and complicate navigation.

---

[7] *Committee of Concerned Journalists: The Principles of Journalism* (2006). Verified April 19, 2017: https://journalistsresource.org/tip-sheets/foundations/principles-of-journalism
[8] Ibid.

"It is what it is," as Robert De Niro says in *Heat*. The world is what it is. But our interpretations may vary severely as a function of political stances, ideological underpinnings, cultural imprimatur, religious convictions, and so forth. This is not tantamount to saying that there are no such things as facts and truth out there, but we may not have always found them yet. That's the reason why we keep asking questions in journalism and science. The world is recalcitrant ever so often; sometimes it doesn't reveal its secrets to us right away, if ever; other times we ask the wrong questions; sometimes facts catch up with us in surprising ways, forcing us to admit we were initially wrong politically, religiously, journalistically, or scientifically. We are forced to change our minds even if we thought we were right at the outset. It's annoying, but we become the wiser on the way. It is what it is.

4.3   Information, Misinformation, and Disinformation

Being informed about something requires having correct, factual information about the subject (Kuklinski et al. 2000). The opposite is not to be uninformed but to be *misinformed*. If you are uninformed about something, you do not necessarily have a belief about what the facts are. You may, like Socrates, at least know that you know nothing. If on the other hand you are *misinformed*, you have factually false convictions that you believe to be true. Misinformation misleads citizens, politicians, and journalists. One may misinform others unintentionally by passing on information that is believed to be true but which turns out to be false. If, on the other hand, the misinformation is intended (as hard as that may be to assess or prove), it is *disinformation* (Søe 2014).

Had the Bush administration itself believed in 2003 that Saddam Hussein commanded functioning weapons of mass destruction ready to be fired, the argument for going to war was a case of misinformation to the public. If, however, the administration *did* know that Hussein had no such things and the misleading was thus intended, it would qualify as a case of disinformation. Intended or not, the consequences were sizeable. A poll from 2015 showed that 41% of Americans still erroneously believed that active weapons of mass destruction had actually been found in Iraq,[9] and not, as it were, but old and poisonous yet useless weapon residues.[10] The same poll explains that 19% of all Americans found it "totally" or "partially" true that Barack Obama was not a legitimate American citizen. Misinformation and disinformation work. All you need to do is mix in the right doses of false claims and twisted stories with a touch of truth to sugarcoat the pill.

4.4   True, False, and Everything in Between

Misinformation is rarely all false. If the misinformation is to have effect, it should not too easily reveal its fraudulence. Misinformation must seem reliable in order to effectively mislead people. Misinformation is therefore often a mixture of something allegedly true; something doubtful, twisted, and undocumented; and downright false information. The composite makes it hard to falsify the misinformation once and for all, as there may just be *something* to it. This makes it easier for the receiver to swallow and to form or consolidate actual convictions and political beliefs related to the information. When disinformation succeeds, the consequence is that the formation of political views and the votes cast are not based on the available facts but rather on dubious premises.

These reflections lead to the below scale of information quality, where true statements and a variety of misleading

---

[9] *Public Mind Poll* (2015). "Ignorance, Partisanship Drive False Beliefs about Obama, Iraq." Verified May 10, 2017: http://publicmind.fdu.edu/2015/false/

[10] Chivers, C.J. (2014). "The Secret Casualties of Iraq's Abandoned Chemical Weapons," *The New York Times*, October 14, 2014. Verified June 10, 2017: https://www.nytimes.com/interactive/2014/10/14/world/middleeast/us-casualties-of-iraq-chemical-weapons.html?_r=1

Scale of Information Quality

ZONE 1	
True statements	Verified facts
ZONE 2	
Doctored statements	Framing, exaggeration, omission, cherry-picking of facts
Undocumented statements	Rumors (maybe true, maybe false)
ZONE 3	
False statements	Misrepresentation of, and contrary to, the facts
Lies	
	Intended false statements
Bullshit	Misrepresentation of motives and purpose, pretence, feigning, dissolution of the dividing line between true and false
Fake news	
	Feigned news, misrepresentation of motives and purpose simulating journalism and truthfulness

Fig. 4.3. The further down the scale from zones 1–3, the greater the misinformation.

statements and strategies to undermine the truth are at opposite ends, with all the shades of gray in between (Fig. 4.3).

4.5   Exaggerations, Omissions, and Cherry Picking

Even though there are no direct false claims involved in a biased presentation of a case, it may result in a twisted perception of reality. Setting the agenda (see Chap. 2) is not only about *which* cases get attention but also *how* they are framed and presented. In attention economics, framing may be

F‍IG. 4.4. Fox frames two identical news stories differently for two different audience segments. For the Spanish-speaking segment, the more neutral word "undocumented" is used to describe the student, while the main channel Fox News uses the negative term "illegals". (Parker, J. (2014). "Fox News and Fox News Latino Cover the Same Story, Hilarity Ensues,"*Addicting Info*, August 8, 2014. Verified June 10, 2017: http://addictinginfo.com/2014/08/08/fox-news-and-fox-news-latino-cover-the-same-story-hilarity-ensues/?fb_comment_id= 593132244141420_593734977414480#f2dbfa4a983dc36).

understood as a question of which *aspects* of a given case that get attention, at the cost of others (Fig. 4.4).

The same goes for the way in which a news story is presented. If, say, the topic is unemployment, the reporter may choose to contact an unemployed person talking about all job applications sent in vain. The reporter may on the other hand also reach out to an employer who is having trouble finding employees. Depending on focus, the same topic may be cast either as a question of lazy and greedy, unemployed "villains" or the systemic conditions whose "victims" are the unemployed. If the framing is massively and systematically leaning in one direction, then the result is an unbalanced coverage and presentation of information that may indeed be misleading. If only one side of things is presented, it may happen at the cost of other perspectives and facts that are part of the whole story and the complete picture. Notwithstanding, it also serves as a problem to *always* insist that there are two sides to everything and that they are necessarily equally compelling. Fifty/fifty journalism may be based on a misperception of the ideals of balance and objectivity (Korsgaard 2017). An extreme case would be to confront a flat Earth believer with a scientist presenting the idea that the sun is at the center of things and treat the two as bona fide equal, cosmological positions. The fact that people have a right to their opinion does not mean that all opinions are equally cogent. Besides, although you have a right to your own opinions, you don't have a right to your own facts Daniel Moynihan once instructed us.

When the angle is exaggerated or the framing extreme, the result may be sins of omission. If you insist on focusing on the approximately 9% of youth in Denmark that have been sentenced at least for violating the criminal code, then you might leave undone the bulk of youth being law-abiding citizens. Or if the majority of stories published about Muslims focus exclusively on the small minority committing crimes and/or supporting a radical or version of Islam, it generates a misleading and nonrepresentative picture of the real world and its risks. It may generate lots of clicks but also a great deal of indignation, fear, and polarization.

Akin to sharpening the frames is cherry picking: You pick exactly the cherries from the tree suiting the case in point but ignore or suppress others. In politics, it means omitting the facts not fitting the program, perspective, or point of view. Facts become something to use or ignore according to needs. It undermines the conditions for evidence-informed politics.

Doctoring of statements may also occur by dint of *misrepresentation*. Only about an hour after Nigel Farage, the leader of the British campaign for leaving the European Union, had proclaimed victory in the Brexit vote, he had to admit to the TV cameras that a crucial number his campaign had been based on was incorrect. The leave campaign had promised to add 350 million pounds to public health services every week

Fɪɢ. 4.5. A Brexit campaign bus with an election promise that could not come to pass.

if the British exit came to pass. Farage rejected his own campaign promise as a "mistake" when confronted with it. It was a distortion of the truth to talk about this as a mere "mistake" when, really, even campaign busses in the UK had references on them to this political promise that could not be realized (Fig. 4.5).

4.6   Rumors, Belief Echoes, and Fact Checking

Undocumented rumors may be true or false. Sometimes, there is something to them and sometimes not. Nonetheless, rumors may have an enormous influence on people's perception of a situation or a person, including a politician. The same goes *even if the rumors have been debunked exactly as make-believe*. That makes rumor mongering and smearing campaigns efficient misinformation tactics carried into the world by *belief echoes*.

The term *belief echo* applies to the phenomenon that, even after an alleged political scandal or rumor has been put to rest as fraudulent and established to be false, the belief echo still has an influence on people's perception of the person at the center of the rumor or the scandal (Thorson 2016). An unfounded, negative rumor about a politician which has been fact checked and falsified beyond any reasonable doubt may still damage the politician's name and fame. Belief echoes unfortunately show how fact checking has limited effect or perhaps sometimes even make matters worse. Fact check needs to reiterate the false claim, which in and by itself makes the belief echoes stronger. And even if the fact check is taken at face value, the rumor still damages the reputation of its subjects.

The US fact checking website Politifact run by the *Tampa Bay Times* has won a Pulitzer Prize for its work. They keep an eye on how different politicians and the media do when it comes to dealing with the truth (Fig. 4.6).

Undoubtedly, such reckoning must adjust for political bias, the media outlet's editorial policies and other journalistic inclinations, the reliability of sources, and other contingencies. Even so, the reckoning is rather thought-provoking. If you compare Clinton and Trump during the election season in 2016, Clinton was ahead when it came to the truthfulness of her statements (Figs. 4.7a and 4.7b), even though scoring more than a third as false statements is hardly impressive (Fig. 4.8).

CONSERVATIVE DAILY POST

"FBI confirms evidence of huge underground Clinton sex network."

— *PolitiFact National* on Friday, November 4th, 2016

Ridiculous without evidence

FIG. 4.6. Politifact's assessment of the *Conservative Daily Post*'s story claiming that the Clintons were part of a sex network.

FIG. 4.7a. Hillary Clinton's score board on true/false statements, summed on April 10, 2017, by Politifact.

FIG. 4.7b. Donald Trump's score board on true/false statements, summed on April 10, 2017, by Politifact.

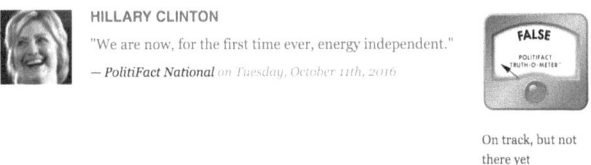

FIG. 4.8. A statement from Hillary Clinton at the nethermost end of the truth value scale.

Obviously, this sort of arithmetic does not mean much if the voters have no faith at all in the fact checkers. A study from September 2016 shows that only 29% of American voters trust fact checks and the media outlets producing them.[11] In a political landscape of distrust, lies may be useful tactics even if unveiled.

4.7   Lies and Bullshit

Exaggerations, misrepresentations, omissions, cherry picking, and rumors may all, alone or combined, comprise bendings of the truth. Yet more undermining are lies, bullshit (Frankfurt 2005), and fake news.

Lies and bullshit are both attempts of deception. There is a difference related to the deceit: the liar tries to deceive by consciously misrepresenting facts. The lie has as its purpose to induce beliefs in the person victim of the lie, thus tampering with subject's perception of the truth.

Bullshit, on the other hand, tries to deceive someone by misrepresenting the sender's real intentions, motives, and purpose. The bullshit's goal is not necessarily to get people to believe the content of what is being said. It may be to make people act in a certain way or to try and get away with something without anyone getting smart about the intention. You may, for instance, bullshit by presenting heaps of irrelevant but factually correct information to take attention away from the real subject. This may furnish the impression that you are trying your best to answer the question, while in reality all you are doing is buying time and avoiding answering.

The liar acknowledges the distinction between true and false but tries to hide the truth. In fact a lie cannot be defined without observing such a distinction. The bullshitter totally ignores the question of truth/false and facts. Bullshit is a bigger threat to truth than lies (Frankfurt 2005: p. 61). Bullshit is of a toxic nature. Stocks of successful bullshit may add to dis-

---

[11] *Rasmussen Reports* (2016): "Voters Don't Trust Media Fact-Checking," September 30. Verified February 5, 2017: http://www.rasmussenreports. com/public_content/politics/general_politics/september_2016/ voters_don_t_trust_media_fact_checking

solving the distinction between true and false. Simulation may have that effect — and bullshit consists in feigning, simulating, or pretending to be or do something other than what you really are or do.

The bullshitter is faking things.[12]

The bullshitter fakes something. That is also what fake news stories do. They fake journalism and news coverage and may be coined "feigned news."

4.8   Fake News as Feigned News

Fake news has been described as "invented material that has been cleverly manipulated so as to come across as reliable, journalistic reporting that may easily be spread online to a large audience that is willing to believe the stories and spread the message."[13]

It must be noted, however, that fake news is not exclusively an online phenomenon. It has existed since long before the invention of the Internet. Neither is fake news inevitably a lie. Fake news can come in the form of an outright lie that tries to portray an intentionally false story as being true. But often, fake news qualifies as being bullshit where the intention is not so much to have people believe an untrue story. Rather, by means of fake news, people are being manipulated to a certain stance or behavior. Even if fake news often consists of false or undocumented claims, distortions, misrepresentations, and so forth, the defining feature of fake news is not the fact that it is bogus. It *simulates* to be journalism and truth-seeking, while its goal is something entirely different. By posing as real news, fake news may pretend to have enlightenment or truth as its end goal, while really it has a

---

[12] Frankfurt (2005: p. 48).

[13] Holan, A.D. (2016). "2016: Lie of the Year: Fake News," *Politifact*, December 13, 2016. Verified April 7, 2017: http://www.politifact.com/truth-o-meter/article/2016/dec/13/2016-lie-year-fake-news/

political or monetary goal aimed for by attention reaping. This goal is kept hidden from the audience. The seeming reliability is enhanced by website addresses that sound real often supplemented with further "testimonials" like pictures and video footage more or less cleverly manipulated.

Four main reasons to foster fake news and put these pseudo journalistic products onto the market may be isolated:

1. Fun/trolling
2. Web traffic/money
3. Marketing/sales
4. Propaganda/power struggle

4.8.1   For Fun/Trolling

"Real news that you can't get from the mainstream media" is the tagline on the news portal, the Underground Report, that deals in feigned news stories such as Michelle Obama having undergone a sex-change operation, Bernie Sanders having Russian connections dating back to the 1960s, and CNN having connections to the Islamic State. The news portal was started on February 21, 2017, by a man named James McDaniel, living in Costa Rica. The website was really a joke or commentary aimed at demonstrating how naïve and gullible Internet users may be. Within a couple of weeks, UndergroundNewsReport.com got more than one million views. McDaniel's fictions, fabrications, and fantasies that catered primarily to a Trump-friendly audience were read and shared widely and received thousands of comments, while only a handful among the audience expressed critical views on the legitimacy of the stories. Three weeks after launching the site, McDaniel made it clear to his readers that is was all a joke. He added: "I was startled that in today's world, so many could be so willfully ignorant. It's truly a frightening time when a group of people screaming, "FAKE NEWS!" at the top of their lungs, live, eat and sleep falsehoods."[14] The website is no longer being updated and stands as a reminder of how easy it is to influence people with fake news.

4.8.2   Web Traffic/Money

A great deal of money may also be made on fake news. Between August and November 2016, "Boris" from Veles in the Eastern European Macedonia earned upward of 16,000 USD on his two websites producing fake news in favor of Trump. Considering that the average wages in Macedonia is 371 dollars a month, Boris decided to drop out of high school in favor of his new occupation of circulating feigned news, even though he, like so many others that partook in this new type of business venture, did not care if Trump won or lost the election. It was merely a question of raising the necessary funds for new cell phones, watches, cars, and drinks at the bar.[15] It was of little interest to the producers of the fake news whether or not the readers believed the stories. Fake news served merely as clickbait. It was digitally produced bullshit with the simple purpose of getting people to invest their click and allocate their attention, which in turn is sold to advertisers. Platforms like Google and Facebook have created strong economic incentives for producing and spreading fake news on this business model. There is money to be made on the market for misinformation. As long as that is the case, it will be difficult to stem the tide of misleading rumors, lies, bullshit, and fake news.

---

[14] Gillin, J. (2017). "Fake News website starts as a joke, gains one million views within 2 weeks," *Politifact*, March 9, 2017. Verified May 2, 2017: http://www.politifact.com/punditfact/article/2017/mar/09/fake-news-website-starts-joke-gains-1-million-view/

[15] Subramanian, S. (2017). "Inside the Macedonian Fake-News Complex," *Wired*, February 15, 2017. Verified April 10, 2017: https://www.wired.com/2017/02/veles-macedonia-fake-news/

Fig. 4.9. A manipulated photo of Angela Merkel supposedly posing for a selfie together with a suspected terrorist in front of Brandenburger Tor in Berlin.

4.8.3   Marketing/Sales

In the market for attention, the actual costumer is the advertising or marketing industry paying for clicks. The marketing industry is quite familiar with the strategy of hiding real intentions by imitating journalistic products. It is called *native advertising* in the industry. The US-American advertising tycoon David Ogilvy (1911–1999), also known under the epithet *The Father of Advertising*, reveals the trick and encourages advertisements to resemble real news stories and journalism:

> It has been found that the less an advertisement looks like an advertisement, and the more it looks like an editorial, the more readers stop, look and read. Therefore, study the graphics used by editors and imitate them. Study the graphics used in advertisements, and avoid them.[16]

The advice has been taken in by the industry, especially in the USA. Ready-made advertisement products dressed up as actual news reports, called *video news release (VNR)*, are

---

[16] Pollitt, C. (2014). "Advertorials in the Age of Content Marketing and Promotion," *Relevance*, November 10th 2014. Verified November 26th 2017: https://www.relevance.com/advertorials-in-the-age-of-content-marketing-and-promotion/

being transmitted undifferentiated during newscasts without being declared as marketing.[17] The same trend may be observed in newspapers containing ads that blend in smoothly with the overall layout of the rest of the paper. This way, marketing is benefitting from and feeding on the credibility that traditionally has been associated with news coverage and journalism. When advertisement is not declared as such, they can rightfully be labeled fake news, even if they do not contain any false statements (Fig. 4.11).

Even though advertisement and marketing industry has lately been making much use of simulated journalism, the industry did not invent fake news. The method of faking journalism is inspired by methods used in war propaganda and power politics. The production of fake news is as old as the printed press. It was originally used as soft power in conflicts, power struggles, and warfare.[18]

4.8.4   Propaganda/Power Struggle

In 1782 during the US-American War of Independence, Benjamin Franklin was in Paris to negotiate peace between England and the USA. During his stay in Paris, he published a fake version of the actual newspaper the *Boston Independent Chronicle*. The fake publication contained a fictional letter that incorporated a false report on the slaughtering and scalping of more than 700 individuals, including peasants, women, children, and infants. The invented incident had supposedly been carried out by the indigenous population who

[17] Farsetta, D. & Price, D. (2006). "Fake TV News: Widespread and Undisclosed." *Center for Media and Democracy*, 16.03.2006. Verified November 26th 2017:

[18] Soll, J. (2016). "The Long and Brutal History of Fake News," *Politico Magazine*, December 18th 2017, verified November 28th 2017: https://www.politico.com/magazine/story/2016/12/fake-news-history-long-violent-214535

Fig. 4.10. An example of the fake news ahead of Kenya's elections 2017. Doctored front page of the Ugandan tabloid Red Pepper on the left, the real one on the right.

acted in accord with the command of the British army.[19] To make the false copy credible, the typography of the actual newspaper was imitated completely with fake ads and announcements (Fig. 4.13).

By means of fake news, Franklin intended to impact public opinion in England and Europe as opposed to merely selling papers to readers and their attention to advertisers. By spreading the story with the help of printed media, the story was envisioned to divide Europe internally swaying the British population together with the broader European public against the war and the King.[20] If successful, the American

[19] Franklin, B. (1782). "Supplement to the Boston Independent Chronicle," [before 22 April 1782], *Founders Online*, National Archives, verifizert   28.11.2017:   http://founders.archives.gov/documents/Franklin/01-37-02-0132

[20] Berry, J. (2017). "The economic efficiency of fake news," *Oxford University Press's Academic Insights for the Thinking World*, January 17th 2017. Verified November 28th 2017: https://blog.oup.com/2017/01/economic-efficiency-fake-news/

position in the war would be strengthened as to the geopolitical power struggle and the specific negotiations for peace.

Dispersing fake news and disinformation as propaganda is about gaining political power and advantages. Advantages entail distrust, internal division and confusion on the side of the opponent, and combativeness, loyalty, and support in your own camp. Fake news can be a soft yet powerful weapon of propaganda in conflicts and wars.

Fake news, partially or wholly constructed stories, false exposure of corruption and fraud or a politician's supposed affiliation to a sinister conspiracy, and so on may entail that the victim of the untruthfulness is considered being corrupt, immoral, or even evil. The same strategy of discrediting and delegitimization may be used against inconvenient journalists and media or even dehumanization whole groups. Disinformation contributes to conflict, polarization, and spiteful feelings potentially threatening to civilized and constructive political debate and social cohesion. *Divide and conquer* is a universal motto motivating opponents on the international geopolitical scene as well as internal agents on both ends of the political spectrum.

The aim of politically motivated fake news is not exclusively to have citizens, journalists, and politicians believe a lie like in Franklin's case above. Mis- and disinformation may be produced and dispersed to confuse or befuddle the public. Russia has been accused of having used precisely this kind of strategy by spreading propaganda domestically and by dispersing disinformation geopolitically. It is contended that Russia is responsible for constructing and putting into circulation conflicting narratives on traditional and social media creating chaos and confusion. In the end, no one can tell truth from falsehoods and the factual from the fabricated.[21]

---

[21] Mariani, M. (2017). "Is Trump's Chaos Tornado a move from the Kremlin's Plauybook?", Vanity Fair, 28.03.2017, verified 17.06.2018: https://www.vanityfair.com/news/2017/03/is-trumps-chaos-a-move-from-the-kremlins-playbook

Regardless of the amount of dis- and misinformation actually originating in Russia and of how well organized it is, it may be said with more certainty that dis- and misinformation *have* caused confusion and disorientation in the USA. According to a survey conducted by the Pew Research Center, 63% of the American respondents answered that the constant stream of fake news during the election created *a great deal of confusion* about what was actually true and what was not. 24% experienced *some confusion*, and 11% were only mildly confused or not confused at all.[22]

When mis- and disinformation reach this level, it is a threat to democracy and security. The international security conference held in Munich 2017 arrives at the same conclusion. The final report with the telling title "Post-Truth, Post-West, Post-Order"[23] identifies the loss of trust in the media and in politicians as the main threat from fake news and disinformation. Mistrust in and delegitimization of journalists and politicians make citizens even more suspicious susceptible to fake news (Colombo and Magri 2017). This undermines trust and legitimacy even more, initializing a vicious circle threatening democracy itself.

Braden R. Allenby, professor at Arizona State University, summarizes aptly how fake news and misleading narratives may be used as weapons on the national as well as the international scene. He defines *weaponized narratives* as "the use of information and communication technologies, services, and tools to create and spread stories intended to subvert and undermine an adversary's institutions, identity, and civilization, and it operates by sowing and exacerbating complexity, confusion, and political and social schisms."[24]

---

[22] Barthel, M, Mitchell, A., & Holcomb, J. (2016). "Many Americans Believe Fake News Is Sowing Confusion," Pew Research Center, 05.12.2016, verified 17.06.2018: http://www.journalism.org/2016/12/15/many-americans-believe-fake-news-is-sowing-confusion/
[23] Munich Security Report (2017), verified 04.04.2017: report2017.securityconference.de/
[24] Allenby, B., R. (2017). "The Age of Weaponized Narrative, or, Where Have You Gone, Walter Cronkite?" *Issues in Science and Technology 33*,

Over the past three decades, the number of women serving time in American prisons has increased more than eightfold.

Today, some 15,000 are held in federal custody and an additional 100,000 are behind bars in local jails. That sustained growth has researchers, former inmates and prison reform advocates calling for women's facilities.

FIG. 4.11. Native advertising that is explicitly declared as such. This ad draws attention to the TV series "Orange Is the New Black." The ad is from *The New York Times* in 2014. Besides containing factually correct information about female inmates and their conditions in prison in the USA, the ad simulates the layout and the style of editorials.

Several of the fake stories that circulated online during the German election in 2017 affirm Allenby's definition of *weaponized narratives*, regardless of the actual intention and origin of the fake stories in question. During the election in Germany as well as during the US-American election, stories were highly polarizing; they fostered suspicion. Those conspiring narratives are a perfect tools for fueling division, subversion, and distrust.

4.8.5 Merkel the Supervillain

In Europe, fake news has also made its way into the media. For instance, stories were run about German Chancellor Angela Merkel not only taking selfies with terrorists (Fig. 4.9) but also being mentally disturbed and simultaneously secretly

---

no. 4, Summer 2017. Verified December 1st 2017: http://issues.org/33-4/the-age-of-weaponized-narrative-or-where-have-you-gone-walter-cronkite/

September 2 at 11 58am

Merkill gratuliert zur Kinderehe. Wie gestört ist diese Person.

Dieses Bild würde Merkel am liebsten verbieten lassen!

Fig. 4.12. Angela Merkel greeting Muslim child brides according to the tweet.

running big German media outlets such as the public broad-casting company ZDF like a puppet master (Fig. 4.11).

Most misinformation centered on Angela Merkel's open door policy of refugees fleeing wars in Syria and other countries in late 2015. According to the UN, no other country in the world received as many requests for asylum during this period as Germany, which ended up welcoming more than a million refugees. This has sparked a backlash against Merkel's policy feeding misinformation predominantly related to refugees, migrants, Muslims, Islam, and rumors about refugees committing crimes or being granted excessive welfare benefits. Some stories are generated locally; some were be tracked back to Russian trolls.[25]

One circulated fake news story that gained some momentum was a picture of Chancellor Merkel flanked by several young women dressed in white. The caption text is "Merkel wünscht den kinderbräuten alles gute" suggesting Merkel wishing Muslim child brides good riddance (Fig. 4.12).

The true story about the picture is that it was taken while Merkel visited a refugee camp in Turkey in April 2016, and the women dressed up in their finest white outfits greeting her welcome—nothing to do with child brides and wedding gowns. But this fact did not stand much of chance at peak of the fairly intense online circulation. Nevertheless, overall it seems that Germany was largely spared much fake news and misinformation during the election.

While Germany got off relatively easy, Kenya was hit hard with fake news campaigns, and scores of misinformation were detected during the election cycle in 2017.[26] According to a survey from Portland and Geo Poll, they found that 90% of the respondents had seen or heard false reports, while 87% of the respondents reporting deliberately false or downright fake news stories (Fig. 4.10).[27]

---

[25] Zeller, F. (2017). "Germany on guard against election hacks, fake news," *Mail & Guardian*, 21.09.2017, verified 17.11.2017: https://mg.co.za/article/2017-09-21-germany-on-guard-against-election-hacks-fake-news

[26] Kamau, A. (2017). "Deaths, defections and deceit: How Kenya's fake news spreads," *African Arguments*, 02.08.2017, verified 15.06.2018: http://africanarguments.org/2017/08/02/deaths-defections-and-deceit-how-kenyas-fake-news-spreads/

[27] Portland and Geo Poll, 19.07.2017, verified 15.06.2018: https://portland-communications.com/pdf/News-Release-The-Reality-of-Fake-News-in-Kenya.pdf

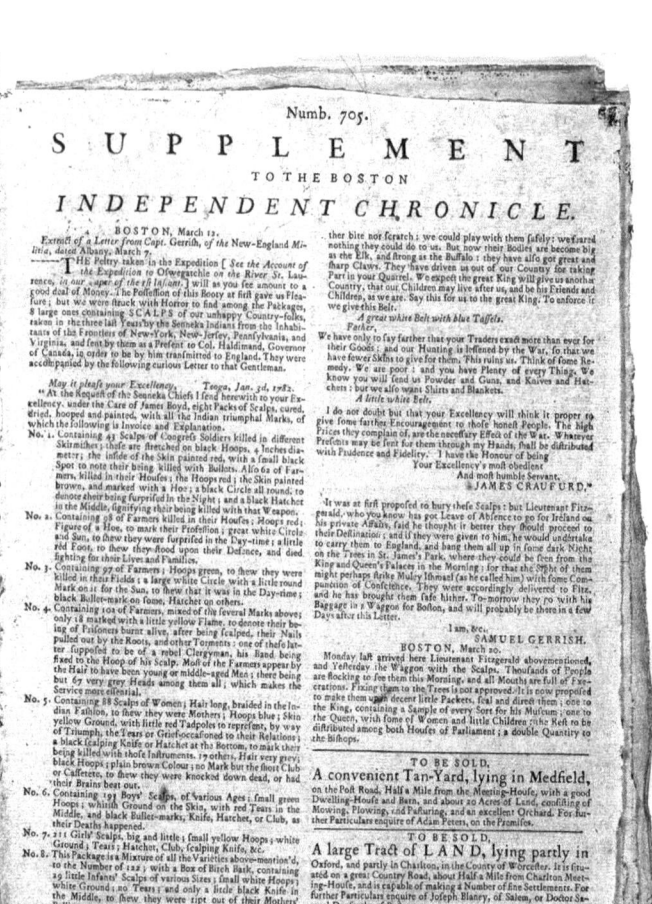

FIG. 4.13. The feigned news story with Benjamin Franklin as its author in the fake newspaper from 1782. (Onion, R. (2015). "The Atrocity Propaganda Ben Franklin Circulated to Sway Public Opinion in America's Favor," *Slate*, 01.07.2015, verified 17.06.2018: http://www.slate.com/blogs/the_vault/2015/07/01/history_of_benjamin_franklin_diplomacy_propaganda_newspaper_with_stories.html).

4.9   Information Whitewashing

The Internet and social media are potent beds for fake news to spread. The Internet gives access to a potentially sizable audience, the investment to get your say is limited, and there is little gatekeeping: great conditions for anything to spread, whether it is true, false, misrepresentation and rumors, lies, or bullshit and feigned news. With befitting software, it is easy and cheap to manipulate pictures digitally. With the emergence of deep fake videos and new methods to manipulate audio, simulating the real thing is only going to get ever more palpable. Add to this the automatized spreading of misinformation through robots. "Bots" are programs that may look like a human Twitter user, for instance, but spread information automatically through the network.

Emilio Ferrera, computer scientist at University of Southern California, estimates that approximately 15% of all Twitter profiles are bots. Among Twitter profiles, the number of fake followers or bots differs considerably. According to a Twitter Audit evaluation from December 2017, 43% of Donald Trump's followers on Twitter (@realDonaldTrump) are fake users being bots.[28] By means of fake followers, the impression may be instilled that the support of a candidate, the anger toward an opponent, or the interest in a news story is bigger than it actually is. A story seems more credible the bigger its circulation and the higher the attention rate it receives. A story being read and shared by many people may work as a *social proof* for the quality of the story and its legitimacy. If the social proof is strong enough to trump critical reflection, you may end up doing what you are doing or reading and believing what you are reading just because you believe others are doing the same (Hendricks and Hansen 2016). Such social psychological *lemming effects* or *information cascades* are bots able to contribute and reinforce.

---

[28]Twitter Audit Report "@realDonaldTrump" 03.12.2017, verified 03.12.2017: https://www.twitteraudit.com/realDonaldTrump

The more misinformation is circulated and repeated and finds its way to new platforms and media houses for both the authorized and the alternative press, the more *whitewashed* the misinformation becomes. There are examples of stories made up for fun or as satire that with time have become whitewashed through circulation and shared as news stories, in order to later be shared or referred to in books as examples of polarized party politics or geopolitical misinformation.

The Internet, digitalization, and social media have created an environment for debate and opinion formation in which news products far from being prime flourish just as well as documented truthful news stories (Moncanu et al. 2015). When it comes to competing for attention, truth does not outweigh false, lies, and bullshit (Vosoughi et al. 2018). Whatever is true is not necessarily viral, and whatever is viral is not necessarily true. In a free and unregulated market for information and news, the inferior news products are not automatically weeded out. In opposition hereto is an overly optimistic rationale of the *marketplace of ideas* in which truthful information will without any interference and censorship outcompete untruthful ideas in public sphere:

> The ideas and opinions compete with one another, and we have the opportunity to test all of them, weighing one against the other. As rational consumers of ideas, we choose the "best" among them. In the same way that "bad" products naturally get pushed out of the market because of the lack of demand for them and "good" products thrive because they satisfy a demand, so also "good" ideas prevail in the marketplace and "bad" ones are weeded out in due course.[29]

A precondition for the ability of the marketplace of ideas to efficiently sort the good information products from the bad ones is the assumption that the consumers of ideas are rational. That they clearheadedly weigh ideas against each other and evaluate according to available evidence as *rational agents* would. Rational agents base opinions on facts and sober reasoning period. But we are not rational agents or

---

[29] Gordon (1997).

exemplars of *homo economicus*. We are humans — and humans are affective beings motivated by emotions more than reason (Hume 1739; Freud 1917; Haidt 2001). This makes us susceptible to all kinds of trickery, deceit, and emotional manipulation as well as resistant to inconvenient facts. Psychological mechanisms are part of the picture. We believe what we want to believe.

Chapter 5
Fact Resistance, Populism, and Conspiracy Theories

5.1 Truthiness

In 2005, the concept *truthiness* was coined by Stephen Colbert, host of the popular satire show, *The Colbert Report*. Truthiness has been referred to as truth that comes from guts and not from facts[1] and is defined as "the belief in what you feel to be true rather than what the facts will support."[2] The concept took hold. In 2006 it was declared word of the year by the Merriam-Webster dictionary. It was used particularly and critically in reference to the political scene of the conservative right in the USA at that time. Before Breitbart, the conservative right rallied around Fox News, whose biased news coverage was satirized by *The Colbert Report*. The show's critical satire focused on how, especially in the conservative right wing and for then President Bush, it often was enough that something felt like it was true in order to be

---

[1] Schlossberg, M. (2014): "One of The Best Moments On 'Colbert Report' Was When He Coined 'Truthiness' In 2005," *Business Insider*, December 18, 2014. Verified June 10, 2017: http://www.businessinsider.com/the-colbert-report-truthiness-clip-2014-12?r=US&IR=T&IR=T

[2] Canfield, D. (2016): "Stephen Colbert Says Oxford Dictionaries' *Post-Truth* Is Just Watered-Down Truthiness," *Slate*, November 18, 2016. Verified June 10, 2017: http://www.slate.com/blogs/browbeat/2016/11/18/watch_stephen_colbert_hit_the_oxford_english_dictionary_for_ripping_off.html

© The Author(s) 2019   79
V. F. Hendricks, M. Vestergaard, *Reality Lost*,
https://doi.org/10.1007/978-3-030-00813-0_5

accepted as such. And not only in the conservative right wing may gut feeling replace truth; this is a universal human phenomenon. The phenomenon of truthiness may find support in cognitive psychology. Through experiments cognitive psychology has demonstrated just how much political bias matters when selecting information and accepting it as true or rejecting it as false.

5.2   Inconvenient Facts

There is a tendency for people's political opinions to be decisive as to what they listen to and believe to be factually sound information.

It feels good to be right and have the facts on your side. Acknowledging that perhaps you are wrong is a different kettle of fish. If you receive factual information that does not fit with your views or downright contradicts them, conflict between this information and your convictions, opinions, and values may result in *cognitive dissonance* (Festinger 1957).

Cognitive dissonance is an unpleasant mental state. One way to avoid it is to be rather selective pertaining to the information to which attention is paid. This is essentially the story of *selection bias*. Selection bias is a tendency, documented through experiments, to pick information and information sources that fit what we want to hear or believe. We pay attention to what we want to be true and avoid inconvenient truths (Manjoo 2008). This is reflected in media consumption. Especially in the USA, there is a strong tendency for people to pick the media whose news coverage fits their political convictions.

A study in selection bias with respect to news sources has documented that more Republicans will read a news story from Fox News than if the source is unknown or, worse still, is from CNN or NPR. The same goes for Democrats: They too pick news stories based on sources, except they have opposite preferences and do not pick their sources quite so markedly (Fig. 5.1).

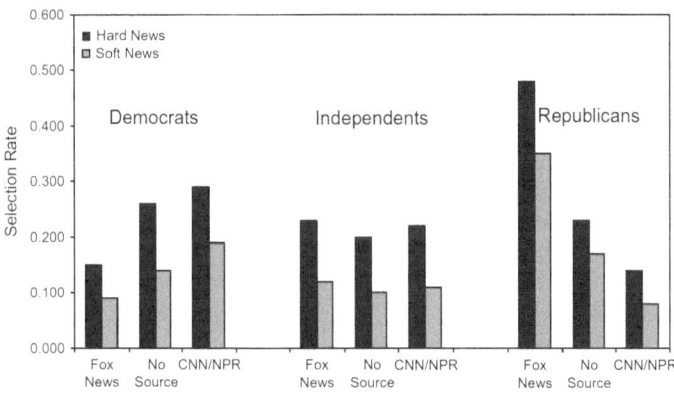

FIG. 5.1. Graph of the choice of news sources distributed according to political affinity. (Source: Iyengar and Hahn 2009).

The human psyche helps making politically colored news coverage a profitable business. Telling people what they want to hear sells news and attracts viewers.

If, in spite of your selection bias regarding choice of information sources, you are unlucky enough to be exposed to information that challenges your ideas with inconvenient facts, there is another phenomenon that may come in handy. You simply write off the inconvenient information that you are presented with as unreliable. That way you avoid the dissonance. This is what is referred to as *motivated reasoning.*

5.3 Motivated Reasoning

Motivated reasoning turns the relationship between ideas and facts on its head. Ideally, you base your ideas and opinions on facts. However, when using motivated reasoning, you start at the other end with a fixed idea and only accept the facts that back it up.

> [Motivated] reasoning starts (...) with the conclusions and works itself back to find the 'facts' that support what we already believe. And if we are presented with facts that contradict our convictions,

we find shrewd ways of rejecting those facts. We are more devious
defense lawyers than we are objective scientists.[3]

Motivated reasoning is a strong element in the distrust of science. It has been revealed that there is a statistical correlation
between believing in an unregulated market economy and
skepticism toward climate research and the reality of anthropogenic climate changes (Lewandowsky et al. 2013). The study
offers the explanation that the claim of CO_2 emissions being a
real and big threat is itself a threat to the market economy that
depends massively on fossil fuels. The principle seems to be
that it is better to distrust the reliability of the climate research
than to revise views of the market. The same tendency to resist
facts may be detected in left-wing politics when it comes to the
impact of weapons legislation.[4] This tendency is so clear that
people's ability to solve math tasks is impaired if the result is
not to their liking. Motivated reasoning must go deep if it even
affects the capacity to calculate. Better to get facts and even
calculation rules out of the way than to change your point of
view and allow facts to disturb your convictions, political identity, and perception of reality. Why would one want to allow
that if one is so convinced of being right?

An experimental study (Kuklinski et al. 2000) has exam-
ined the connection between political opinions and factual
knowledge related to welfare benefits. The study shows that
those people who have the most ideological bolstered opinions are also the ones who have a tendency to be most factually wrong. Notwithstanding, the study also demonstrates that
those very same people are the ones to be most sure of themselves and convinced that they are right. It may thus be difficult to convince those that have the greatest need for a dose
of facts. Given these psychological conditions, misinformation has it easy. And even more so when we become polarized

---

[3] Jones, D. (2016): "Seeing reason: How to change minds in a 'post-fact'
world," *New Scientist*, November 30, 2016. Verified June 10, 2017: https://
www.newscientist.com/article/mg23231020-500-changing-minds-
how-to-trump-delusion-and-restore-the-power-of-facts/
[4] Klein, D. (2014): "How politics makes us stupid," *Vox*, June 4, 2014.
Verified   June   6,   2017:   https://www.vox.com/2014/4/6/5556462/
brain-dead-how-politics-makes-us-stupid

and divided into opposing groups, identifying with our tribe and thinking in terms of *us-versus-them.*

5.4   Loyal Lies

Truth is the first victim in war. The same is the case in a situation of (cold) civil war where the political landscape of a polity is so divided and polarized that the opposing fractions consider each other not just as opponents but as *enemies.* If tribal thinking, or tribalism, in which you identify strongly with a group of which you form part, becomes sufficiently prominent, then politics boil down to a friend or foe relationship in which truth, and often also substance, comes in on a very distant second. Then, it is only about winning. Telling lies and manipulating and spreading disinformation are considered fair game in warfare. If those in the opposing group, *the others*, are seen as an enemy, *blue lies* are legitimate and a way to go too. The expression blue lies is inspired by cases in which police officers have lied out of loyalty to the group to cover for colleagues, or in order to ensure conviction of an indicted person (Barnes 1994). Blue lies are lies on behalf of a group that serve the group.[5] The lies may strengthen the internal coherence of the group and loyalty among its members. Those who are not part of the group, however, pay the price. If the loyalty to the group of police officers is greater than the loyalty to the law and the citizens, it undermines the rule of law; and if political loyalty to a political party, the minister, secretary, cabinet member, or President is greater than loyalty to the law, the constitution, and citizens as such, it undermines democracy. Since politics and population have become so polarized in the USA, blue lies may be part of the reason why so many of Trump's supporters do not seem to react negatively to the revelations of falsehoods and downright lies (Fig. 5.2).

---

[5] Fu, G. et al. (2008): "Lying in the name of the collective good," *PMC*, October 20, 2008. Verified June 10, 2017: https://www.ncbi.nlm.nih.gov/pmc/articles/PMC2570108/#R1

Trump voters strongly approve of his presidency

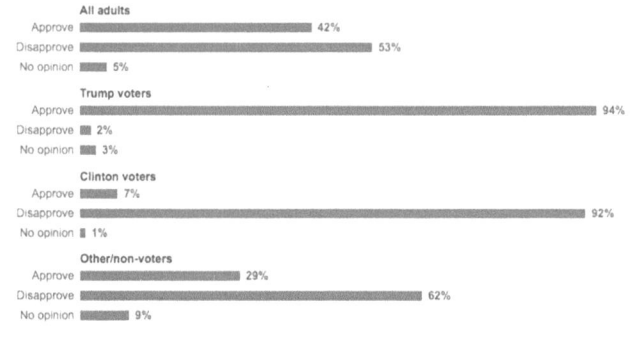

Source: Washington Post/ABC News (America) survey, April 17-20, 2017

FIG. 5.2. Poll from April 17, 2017, shows a deep polarization in the evaluation of Trump as President. 94% of Trump's voters applaud him, and only 2% are dissatisfied. Among Clinton's voters, 7% applaud him, while 92% turn their thumbs down (Hanrahan, C. (2017): "Donald Trump: Is he the most unpopular United States president in history?", *ABC News*, April 28, 2017. Verified June 10, 2017: http://www.abc.net.au/news/2017-04-28/donald-trump-is-he-the-most-unpopular-president-in-history/8469854).

Trump's lies are lies on behalf of "the movement" and in its favor against an enemy who needs to be fought. So says politics researcher George Edwards from Texas A&M University, who explains the lack of reaction from Republicans to the revelations of Trump's falsehoods as a result of tribal thinking, deep polarization, and group internal acceptance of blue lies as legitimate weapons against the others:

> People applaud lying to enemy nations, and since many now view those on the other side of American politics as enemies, they may feel that lies, when they recognize them, are legitimate means in the warfare.[6]

---

[6] Smith, J.A. (2017): "How the Science of 'Blue Lies' May Explain Trump's Support," *Scientific American*, March 24, 2017. Verified June 10, 2017: https://blogs.scientificamerican.com/guest-blog/how-the-science-of-blue-lies-may-explain-trumps-support/

When polarization is so pervasive and tribal thinking so prominent, they compromise the capacity for observation in the first place and then also the willingness to report truthfully what is observed. The factual question as to which crowd is the bigger one in two photos where the difference is clear (see Chap. 4) may under these circumstances become a political question, the answer to which depends on political affinity (Fig. 5.3).

Psychological bias phenomena and social psychological group and polarization dynamics not only cause fact resistance; they also contribute to popularizing, simple, identity-building narratives of us-versus-them. In addition to the

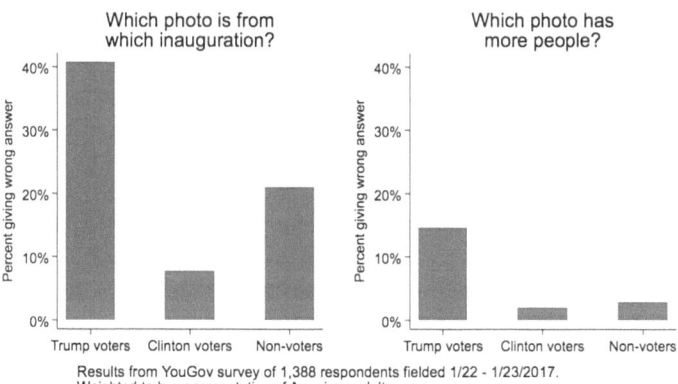

Results from YouGov survey of 1,388 respondents fielded 1/22 - 1/23/2017.
Weighted to be representative of American adults.

FIG. 5.3. Tribal thinking and polarization may cause strong fact resistance. 40% of Trump's supporters believe that the photo from Obama's inauguration on January 9, 2009 (the one with the bigger crowd on the left), was from Trump's inauguration on January 17, 2017. Worse yet, 15 % of Trump's supporters were even willing to claim (perhaps as a blue lie) that the crowd shown on the photo from Trump's inauguration was bigger than the crowd on the photo from Obama's inauguration back in 2009 (Schaffner and Luks (2017): "This is what Trump voters said when asked to compare his inauguration crowd with Obama's," *Washington Post*, January 25, 2017. Verified April 28, 2017: https://www.washingtonpost.com/news/monkey-cage/wp/2017/01/25/we-asked-people-which-inauguration-crowd-was-bigger-heres-what-they-said/?utm_term=.182e2c9af76a).

media environment, the human psyche is as designed for the populism that is doing so well at the moment.

5.5   Populism: Us-Versus-Them

Populism is not a specific political ideology, but rather a *strategy* whose core is the division of us-versus-them. Populism feeds off polarized and excluding narratives about friends and enemies. According to the German historian of ideas and professor of politics, Jan Werner-Müller, the core narrative in populism is that the populists themselves, and only they, represent the true will of the people, singular (Müller 2016). Take the French Front National's slogan "au nom de peuple," *in the name of the people*, or Nigel Farage from the British UKIP's (UK Independence Party) talk of Brexit as a victory for *real people*. Populism divides the population into the real people and the others. It also disjoins politicians into those who represent the "real" people (the populists themselves) and other politicians consequently not representing the people's will. Populists make a symbolic construction of a group they name *the people,* which is identified as exactly their own supporters and constituency. They claim to not only represent 99% of the people but rather all of the people, since the rest are excluded as the "others." Populism is anti-pluralistic.

Even though right-wing populism prospers the most these days, populism is not right-wing in and of itself. As an example, Hugo Chávez, Venezuela's former president, was a left-wing populist of our times. Before his death in 2013, he had run the country close to ruin both economically and democratically in the name of the people, socialism, and democracy. His opponents were proclaimed enemies of the people as well as of democracy.[7] Turkish President Recep Tayyip

---

[7] Fisher, M. & Taub, A. (2017): "How Does Populism Turn Authoritarian? Venezuela Is a Case in Point," *New York Times*, April 1, 2017. Verified June 10, 2017: https://www.nytimes.com/2017/04/01/world/americas/venezuela-populism-authoritarianism.html

Donald J. Trump
@realDonaldTrump

Follow

The FAKE NEWS media (failing @nytimes, @NBCNews,
@ABC, @CBS, @CNN) is not my enemy, it is the enemy of the
American People!

10:48 PM - 17 Feb 2017

51,226   162,120

Fɪɢ. 5.4. A tweet from Donald Trump in which the established
media are presented as the enemy of the American people.

Erdogan follows the same populist recipe in Turkey based on
a religious and Islamist ideology: "We are the people. Who
are you?"[8] If you are not with me, you are against the people,
is the populist refrain. In populist understanding, political
opponents are not seen as representatives of other legitimate
opinions and viewpoints, which is a basic condition for a plu-
ralistic, liberal democracy. Instead, the political opponents are
presented as part of an elite systematically betraying the
people, neglecting their wishes, and paying them no heed
whatsoever. The opponents easily become the other, the
enemy. The others may be immigrants who pose a threat to
national security, identity, or values, according to the populists, or they may be the political elite in Washington, Paris,
Berlin, or Brussels who betray the people by leaving the bor-
ders open and thereby selling out the nation and its legitimate citizens. The others may also encompass the media elite
(and the fact checkers, for that matter) who are routinely
accused of hiding the truth with the intention to silence the
voice of the people and of producing fake news, when the
news coverage does fair well with the populists in question.

---

[8] Müller, J.W. (2016): "Trump, Erdoğan, Farage: The attractions of populism for politicians, the dangers for democracy," *The Guardian*, October
02, 2016. Verified June 10, 2017: https://www.theguardian.com/
books/2016/sep/02/trump-erdogan-farage-the-attractions-of-
populism-for-politicians-the-dangers-for-democracy

In Germany, *Pegida*, a group on the outer right, has been explicitly utilizing the historically charged term *lying press (Lügenpresse)* when referring to the media. The political party *Alternative für Deutschland* (Alternative for Germany) has been more moderate in addressing the press, yet without failing to deliver the same message by referring to the press as *Pinocchio Press* (Fig. 5.4).[9]

5.6   Social Transmission: Indignation and Fear

Populist, excluding, and polarizing narratives of us-versus-them are fit to attract attention and set the agenda in the media. Populism is an efficient media strategy that plays on emotions. The narrative structure of us-versus-them, with the others being villains, is efficient when it comes to mobilizing anger or fear. News stories that provoke anger (i.e., indignation)[10] and fear have a much greater tendency to go viral and suck attention on social media (Berger and Milkman 2012). Negative emotions such as anger and fear and positive ones such as awe and fascination are called *activity mobilizing emotions*. They motivate people to act. These are opposed to sadness, or being comfortable, which are called *activity demobilizing emotions*. If a candidate's statements sadden you, this may demobilize you so you do not vote even though you find the alternative candidate much worse. Acting also means to share, retweet, like, and make other online gestures that fuel the social transmission of media material. If you

---

[9] Rohbohm, H. (2015). »Petry schwört AfD auf „harten Kampf" ein«, *Junge Freiheit*, November 28th 2015. Verified November 25th 2017: https://jungefreiheit.de/politik/deutschland/2015/petry-schwoert-afd-auf-harten-kampf-ein/

[10] In the study Berger uses the concept category of anger, but the news articles categorized thus are rather about indignation. Indignation is anger about what seems to be unfair, as seen in these headlines: "What Red Ink? Wall Street Paid Hefty Bonuses," "Loan Titans Paid McCain Adviser Nearly $2 Million."

want content to go viral, make people red-hot angry or scared. Populist narratives are seldomly totally cut off from reality. Citizens who vote for populists may be both indignant and fearful for good reason given tough and harsh facts they encounter. Rising inequality, social and cultural marginalization, and the rising difference between rural and urban communities, to mention a few, may all be factual reasons for anger and fear. Isolated facts and news may be cherry-picked to support a populist cause. This is one of populism's hallmarks: It simplifies cases and circumstances, cherry-picking facts, and framing topics as well as information according to the stereotypical opposition between us-versus-them. If facts run counter to that core narrative, they may be left out or reasoned away as not valid, with evasive explanations such as: "Even though statistics do not show a rise in crime, there has to be a rise. We simply cannot see it, so it must be a question of shadow numbers." Though shadow numbers do exist, and not everything gets reported to the police, automatically rejecting the numbers whenever they contradict your political agenda shows a basic distrust or disregard for evidence that may undermine a political debate carried out on a factually informed basis.

With populism, stereotypes often replace facts. Scapegoats and simplified explanations become replacements for the world's complex and often less than transparent cause and effect chains (Dahlgren and Alvares 2016). Populists also have a simplified answer to an existential or religious question that man has been shouting to the heavens as long as religion has existed: Why do I suffer?

5.7 Why Do I Suffer?

The pessimistic German philosopher Arthur Schopenhauer (1778–1860) had an unpleasantly sharp eye for how man is a true master of suffering. According to him, our whole life is suffering in different forms (Schopenhauer 1966: pp. 473–479). We may suffer in many ways: from basic physical pain and agony, thirst and hunger, sickness, and poverty to lack of recognition, fear, alienation, rootlessness, social or political marginalization, and stigmatization, to name a few. Suffering takes many forms, and the question "Why do I suffer?" is a basic, existential question that has been posed time and again, ever since Job did it in the "Book of Job" in the *Old Testament*.

The story of Job is a story of suffering. The innocent Job is hit by one catastrophe after another because God and the Devil have made a bet as to whether Job will stick to his belief in God regardless of massive misfortune. Job's friends are not exactly great friends; they insist that Job must have sinned in order to be hit by such suffering. As they see it, all suffering is God's punishment for sinning, and since God is justice incarnate, the punishment must be just: "…those who plow evil and those who sow trouble reap it" (Job: 4,8). They therefore believe that Job must necessarily be guilty but might have forgotten it himself, because you reap what you sow.

Until modern times, this was religion's standard response: Have you been hit by the plague? Surely, you must have sinned. An earthquake? God's punishment! That will teach you. This is theological *theodicy* tradition, which defends God as good and almighty in spite of the existence of suffering. The term itself, *theodicy*, is a compound of *teo* = God and *diké* = justice. It was introduced by the philosopher Gottfried Wilhelm Leibniz (1646–1716), who added the philosophical remark that suffering is a necessary evil in the best of all possible worlds[11] to the traditional religious explanation of suffering as punishment. Supposedly, the reason man cannot see this is due to our limited perspective; but God sees it, and the Lord works in mysterious ways.

There are examples of more modern responses in classic literature. The French philosopher Voltaire's book *Candide* is a satirical debunking of the perception of suffering as deserved or a necessary evil (Voltaire 1991). In Voltaire's

---

[11] *Theodicy* by Freiherr von Gottfried Wilhelm Leibniz, verified June 10, 2017: http://www.gutenberg.org/files/17147/17147-h/17147-h.htm

opinion, it adds insult to injury to preach that it is people's own fault or that suffering is a necessary ingredient in the best of all possible worlds. He rejects the idea that it does not suffice to suffer but rather posits that one must also endure additional suffering from knowing that one has brought it upon oneself or accept one's own and other people's suffering as building blocks in universal harmony. Russian author Fyodor Dostoyevsky's character Ivan Karamasov likewise rejects salvation for that exact reason: If the suffering, particularly that of children, is necessary for the salvation and the harmony of the whole, then he feels the ticket to Paradise comes at too high a price and says no thanks (Dostoyevsky 2002).

Even though the great religious theodicy explanations have gone out of fashion, without however disappearing from fundamentalist circles, the explanatory scaffold has lived on. It has been secularized and has found its way into politics and economics. The narratives that are being served to people stuck at the lower end of the income scale, or who have lost their jobs as a result of globalization, rationalization, and automation, are akin to the traditional theodicy explanations in their basic structure. Here are a few examples taken to their absurd extreme:

1. You suffer (say by losing your job) because:

    (a) You fail to be flexible enough for the globalized market.
    (b) You have no useful training or education.
    (c) You are part of the "basket of deplorables,"[12] which was how Hillary Clinton referred to a large part of Trump's electoral base.

---

[12] At a fundraiser in New York on September 9, 2016, during the election campaign, Hillary Clinton said as follows: "You know, to just be grossly generalistic, you could put half of Trump's supporters into what I call the basket of deplorables. Right? They're racist, sexist, homophobic, xenophobic — Islamophobic — you name it."

And this is why you largely deserve the situation that you find yourself in: You reap what you sow.

Even if you are one of the people who lost their jobs, globalization is generally a good thing, since it is good for the economy and growth in the long run. Your unemployment is a necessary evil in the best of all possible economic systems, which is why things only look dark from your limited perspective. In reality, seen from the view of the whole, it is for the common good in the long run and will create growth, jobs, and progress.

In the best of all possible economic systems the individual is rewarded according to its contribution and merits. The market is rational because the reward is proportional to effort: the income equals productivity which equals benefit for society. The wealthiest as well as the individuals placed at the bottom of the income scale harvest as they have sown. The growing inequality and the wage stagnation for the middle class are within this reasoning nothing else but a manifestation of a higher justice, of "just deserts" (Mankiw 2010: 17).

Fɪɢ. 5.5. The Trump badge with the text "deplorable lives matter" plays both on Hillary Clinton's labeling of Trump supporters as deplorable and on the movement Black Lives Matter, which was formed in 2012 in protest against controversial police killings of African-Americans in the USA.

If these are the theodicy explanations that people in general have been provided from politicians and (neoclassical) economists, small wonder that they turn to populist narratives that at least do not belittle their suffering by depicting it as an illusion or even deserved (Fig. 5.5).

Populism employs a secularized version of the myth of the fall of man to explain suffering as something more palatable to the sufferer. Things have gone wrong, suffering has come into the world with the others (the immigrants, the political elite, the established media), and what we need to do now is return to the paradisiacal state that existed before the fall, i.e., "Make America Great Again!"

The populist narratives, as opposed to the theodicy narratives, have a simple and comfortable answer as to why we suffer: There is a scapegoat, a culprit to blame for most or all evil, and an implicit vow to remedy the suffering.

Q: Why do I suffer?

A: You suffer because of *them*! Because of the others.
     And we, the populists, aim to do something about that!

This answer lends populist narratives an existential strength that is hard to beat. Nevertheless, their simplification of complex matters by means of a scapegoat entails that they lose touch with reality and cannot honor their promise of the imminent return to Paradise, at least not with anything other than a symbolic and elusive feeling of change for the better.

> The main focus of populist policy is therefore to tend to these people waiting—to give them a reason for their suffering, to verbally recreate the post-factual world of their beliefs, to make them feel like they are moving forward. Populism is not a system of facts or solutions, operating in the complex world of policy and legislation, but rather an interactive fiction, borne of posturing and symbolism, where whole countries can become not what they are, but what they believe themselves to be.[13]

---

[13] Rondón, A.G. (2017):" Donald Trump's Fictional America," *Politico Magazine*, April 2, 2017. Verified April 26, 2017: http://www.politico.com/magazine/story/2017/04/donald-trumps-fictional-america-post-fact-venezuela-214973

When the us-versus-them structure gets further radicalized, it ends up in conspiracy narratives in which the others, the elite who wish us ill, operate in the dark and run the whole show secretly at a deep state level.

5.8   The Structure of Conspiracy Theories

Populist narratives may go into overdrive and become conspirational. If the other politicians and not least the journalists do not work for the people, then whom do they work for? A foreign power or a secret elite? Conspiracy theorists turn up the distrust of the establishment, which could be run by the enemy, the others, and they use this distrust as fuel. Thus, conspiracy stories and theories have the same basic us-versus-them structure as the populist narratives, but taken to an even more extreme degree. The others are hidden in conspiracy thinking, and there is also an even greater tendency to perceive the world unnuanced as a battlefield for an epic battle between good and evil that only those in the know can see. A conspirational conviction may be defined as being a belief that an organization consisting of individuals or groups is plotting and acting in the dark in order to reach a specific goal that is often malignant.

Conspiracy theories are not necessarily wrong. Watergate is a prime example of a real conspiracy, and there are also more or less shady secret intelligence missions, corrupt actions, clandestine operations, and unsavory political deals that sometimes come to light but oftentimes go unnoticed.

The belief in secret agendas may go into hyperdrive and become a paranoid perception of the world in which everything, give and take, are perceived as being run by the others. This turns critical thinking into conspiracy thinking, where motivated reasoning and fact resistance may thrive and misinformation is accepted uncritically as long as blame is attributed to the preferred quarters. An encompassing study looked into 2.3 million Facebook users' information consumption and established that people with conspiracy convic-

Louise Mensch ⬮
@LouiseMensch

🐦 Follow

That's because you, Russia, funded riots in Ferguson. See 0
hour I have your connections to Trump archived via Schiller and
Scavino twitter.com/repfunder/stat…

2.08 PM · 9 Apr 2017

↩   ⟲ 226   ♥ 277   ℹ

FIG. 5.6. Conspirational paranoia in the left wing where undocumented claims and fake news regarding the Kremlin-gate scandal thrive. Here is an accusation that Russia is behind riots in Ferguson.

tions have a greater than average tendency to accept fake news and undocumented claims (Mocanua et al. 2015). When a critical sense turns into conspirational thinking, alternative facts are swallowed hook, line, and sinker, as long as they come from alternative sources according to one's taste and world view. This does not only go for the American right wing, although it continues to be repleted with conspiracy tendencies. Ever since the US Presidential Election 2016, factions of the American left wing have gone into conspirational mode and seen Russian infiltration here, there, and perhaps everywhere. The election result is not the only thing explained as the result of Russian meddling; Russia is even ascribed responsibility for the police violence that occurred in Ferguson (Fig. 5.6).

The tendency to bring and share undocumented claims and fake stories uncritically, combined with distrust of the "mainstream media," seems to be part and parcel of conspiracy theories.[14]

Three principles are characteristic of conspirational thinking (Barkun 2013). Firstly, nothing happens by chance; there is always an (evil) will or intention behind it. Secondly, what-

[14] Beauchamp, Z. (2017). "Democrats are falling for fake news about Russia," *Vox*, May 19, 2017. Verified June 10, 2017: https://www.vox.com/world/2017/5/19/15561842/trump-russia-louise-mensch

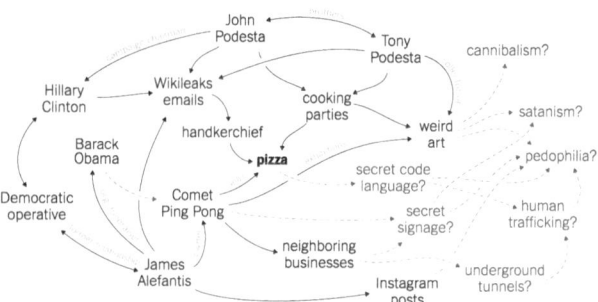

Dissecting the #PizzaGate Conspiracy Theories

By GREGOR AISCH, JON HUANG and CECILIA KANG   DEC. 10, 2016

FIG. 5.7. The conspiracy theory #Pizzagate regarding purported sexual abuse of children in the cellar beneath Comet Ping Pong pizza restaurant in Washington, D.C., where there actually is no cellar at all, managed to connect everything from handkerchiefs to Barack Obama, Hillary Clinton and weird art to human trafficking, pedophilia to Satanism and cannibalism. (Aish, G. and Huang, J. (2016): "Dissecting the #PizzaGate Conspiracy Theories," *The New York Times*, December 10, 2016. Verified December 17, 2016: http://www. nytimes.com/interactive/2016/12/10/business/media/pizzagate. html?_r=0).

ever happens must be connected to the rest, and that includes all news and facts that come to light. All of it forms part of a narrative in which everything fits, makes sense, and comes together in a coherent world view that often, however, loses its connection to reality due to its simplistic divide of the world into good or evil. If everything fits into a theory, most likely the theory itself does not square with reality (Fig. 5.7).

When everything is connected, and nothing happens by accident, then there are hidden patterns behind it all that explain everything. The task of the conspiracy theorist is to connect the dots between facts, rumors, and fake news with lines that reveal the entire wicked plan (Fig. 5.8).

Thirdly, according to the conspiracy theories, nothing is as it seems, and the official story is certainly untrue. The motto

Fig. 5.8. If you want to see a monster, all you have to do is connect the dots, and you will see the outlines of it.

The concept of global warming was created by and for the Chinese in order to make U.S. manufacturing non-competitive.

24,831   14,654

2:15 PM · 6 Nov 2012

FIG. 5.9. Trump's tweet from 2012 in which he declared man-made climate changes to be a Chinese hoax made up to destroy America's competitiveness.

from the TV series *X-files* sums it up: "Trust no one!" This thinking leads to enormous distrust in the common and established knowledge-producing and fact-checking bodies, among them the mainstream media, the educational system, and the science and research institutions. The usual channels for information are thought to be filtered and controlled by the plot or active participants in it. In this way, every piece of information that questions the conspiracy theory may be written off as being planted by those that conspire to mislead you. This is motivated reasoning run wild. The heavyweight conspiracy theories cannot be falsified, exactly because any attempt at falsifying them is perceived as a trap set by the others.

   The more global and all-encompassing the conspiracy theory, the harder it is to shoot it down, but the more unrealistic it is as well. The heavyweight theories concerning super conspiracies, where *everything* is connected and controlled by a secret world elite or deep state, ascribe humans unrealistic capacities to make long-term, secret plans, carry them out successfully, and resist to brag about it. Some of the heavyweight theories turn into a sort of occultism in which those who control the world are not humans at all, but demigods, reptiles, aliens, or other supernatural beings. It takes supernatural powers to keep track of so much, have so many allies, and keep everything a secret.

One middleweight conspiracy theory is the one claiming that anthropogenic climate changes are a Chinese hoax. It is a cover for a shady plan to undermine American market competitiveness. This theory was proposed by Trump in 2012 (Fig. 5.9). It would require almost supernatural skills to carry out such a plot with that many fellow conspirators, among them almost all the climate scientists on Earth.

The argument Trump used 5 years down the line, on May 31, 2017, when he withdrew the USA from the Paris Agreement, the United Nations accord that aims to lower greenhouse gas emissions and mitigate climate change impacts, is composed out of the same basic conspirational melody. There is a suspicion of the others having shady intentions of undermining the American economy. These intentions differ from the official reason given, which is to mitigate the rise in temperature:

> This agreement is less about the climate and more about other countries gaining a financial advantage over the United States. The rest of the world applauded when we signed the Paris Agreement — they went wild; they were so happy — for the simple reason that it put our country, the United States of America, which we all love, at a very, very big economic disadvantage. A cynic would say the obvious reason for economic competitors and their wish to see us remain in the agreement is so that we continue to suffer this self-inflicted major economic wound.[15]

Apparently, the Paris Agreement is not about the climate at all. It is only a front, a shady official story that covers up villainous intentions of hitting the American economy. And it is a plan in which the whole world must have ganged up to carry out. If this was actually the case, then NASA would be part of the conspiracy. NASA describes man-made climate changes as a fact beyond any reasonable doubt: The evidence is too

---

[15] Read: Trump's speech announcing withdrawal from the Paris Agreement on climate change, *CNN*, June 01, 2017. Verified June 10, 2017: http://edition.cnn.com/2017/06/01/politics/trump-paris-agreement-speech/index.html

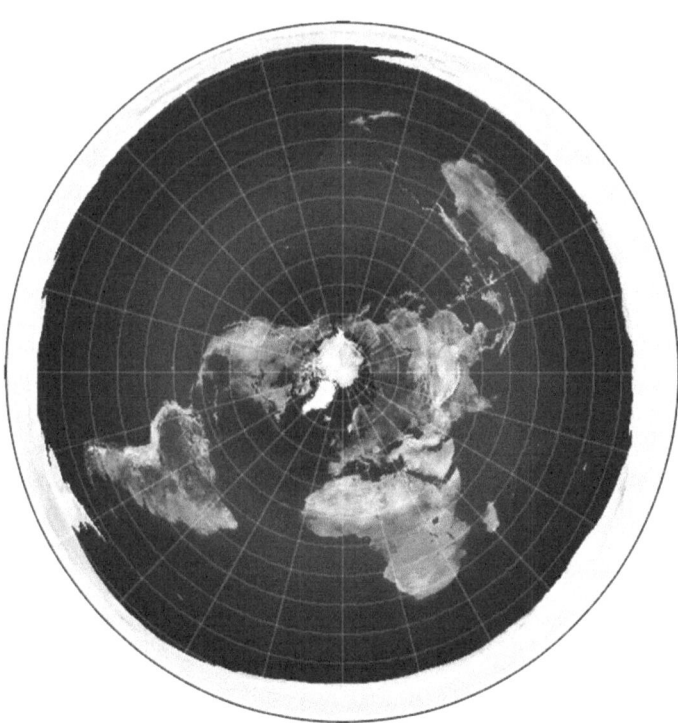

FIG. 5.10. Flat Earth Society's world map. Antarctica is not depicted as the southernmost continent, containing the south pole, but instead as a ring with a tall and impassable wall surrounding the flat Earth disc.

strong.[16] But wait a minute. Was it not NASA who faked the moonlanding in a film studio? How can we trust them on anything?

If distrust in the knowledge institutions grows deep enough, it may end in a skepticism so comprehensive that even the fact that the Earth is a globe is not a given. This is the starting point for a growing movement called Flat Earth

---

[16]"Climate change: How do we know?," *NASA*, verified June 10, 2017: https://climate.nasa.gov/evidence/ ᴵ

Society that insists the Earth is not a globe and argues that people think so erroneously due to the lies and photoshopped globe propaganda from a big villain, NASA. NASA is thought to have bribed every single astronaut to go along with the hoax and lie to the public (Fig. 5.10).[17]

On this view, citizens, politicians, scientists, and teachers who believe the Earth is round are either misled or themselves consciously misleading the public. If you cannot trust anyone, then what you may know or come to know is very limited. Have you seen with your own eyes that the Earth is round? No, you haven't. So how can you know for sure that it is? If suspicion and distrust metamorphose and win over people beyond small circles, it not only undermines science, enlightenment, efficient political action, and problem-solving. It undermines democracy itself.

---

[17]The Flat Earth Society (2016), verified June 10, 2017: http://www.the-flatearthsociety.org/home/index.php/faq

Chapter 6
The Post-factual Democracy

6.1   The Word of the Year 2016

In 2016 "post-truth" was the word of the year in *Oxford Dictionaries*. Post-truth is defined as "relating to or denoting circumstances in which objective facts are less influential in shaping public opinion than appeals to emotion and personal belief."[1] Oxford attributed the nomination to the fact that post-truth went from a peripheral concept to exploding in popularity in 2016 pace the British vote on the EU leading to Brexit and with the American presidential election.

Post-truth also reached Davos and the agenda of the World Economic Forum. The forum's considering "misinformation" a global risk back in 2013 was followed up 4 years later in their *Global Risks Report* warning that post-truth political debate undermines the efficiency and legitimacy of democracies.[2] Democracy itself, as well as the political capacity to efficiently address and solve social problems, including the global challenges facing the world, is threatened by political debate in which facts matter less than emotions and opinions.

---

[1] Oxford Dictionaries (2017): "Post-truth," Oxford: Oxford University Press. Verified February 4, 2017: https://en.oxforddictionaries.com/definition/post-truth

[2] World Economic Forum: *Global Risks Report* 2017: 23. Verified June 11, 2017: https://www.weforum.org/reports/the-global-risks-report-2017

© The Author(s) 2019
V. F. Hendricks, M. Vestergaard, *Reality Lost*,
https://doi.org/10.1007/978-3-030-00813-0_6

6.2   Post-factual Democracy

Post-factual democracy points to the same phenomenon as post-truth politics: the tendency for facts obtained and verified by reliable methods to play second fiddle or worse in politics. To rehearse: *A democracy is in a **post-factual state** when politically opportune but factually misleading narratives form the basis for political debate, decision, and legislation.*

The "factually misleading narratives" may be lies and tall tales; false, fake, or distorted news stories; or populist or conspiracy us-versus-them narratives with cherry picking or strong framing of facts to support the narratives. When facts are cherry-picked according to their political convenience or facts are replaced by false alternatives or simply denied, they lose their authority as the basis for discussion, debate, and decision. Then facts are reduced to strategic armaments in a political power struggle and are employed or deployed, regarded or disregarded, and accepted or denied according to tactical and strategical needs (Fig. 6.1).

The phrase "sometimes we (The White House) may disagree with facts" uttered by former White House Press Secretary Sean Spicer while debating crowd size and the

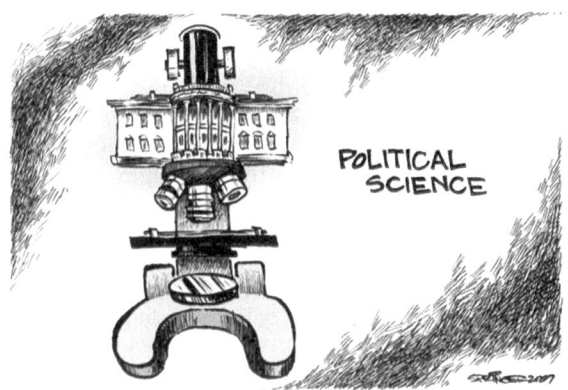

FIG. 6.1. When facts are politicized, reliable inquiry is undermined by political interests.

weather during the Presidential Inauguration 2017 illustrates a selective perception of facts. Such an approach relegates facts to political instruments in debate rather than being the common foundation securing and qualifying deliberation. If verified facts obtained by reliable methods become politicized and reduced to partisan contributions, political debate loses its anchorage in reality. In extreme cases, even the question as to whether the sun shines or not becomes a question, the answer to which depends on your political point of view. In that rabbit hole, everything is relative.

George W. Bush infamously noted after the 9/11 terror attacks in 2001 that in the war on terror all countries must chose side: "You're either with us, or against us." Neutrality was out of the question. In a post-factual democracy, the same principle seems to go for facts and the institutions uncovering and handling them. Everything is political. On the battlefield, all standpoints are perceived as, and suspected to be, nothing but veiled political interest. Neutrality is not an option. If you try to stay neutral, you still risk becoming cannon fodder. Science, journalism, and law are politicized and categorized as friends or foes. You are either with us or against us, and if you are against us, then you are fake news.

In a post-factual democracy, respect for and acknowledgment of the real has disappeared in the heat of battle. Reality, or rather what counts as real, is produced and constructed by those who have the power to do so.

Journalist Ron Suskind has described a situation dating back to 2002 in which he spoke to an adviser of George W. Bush later identified as neoconservative Karl Rove giving voice to post-factual politics:

> The aide said that guys like me were "in what we call the reality-based community," which he defined as people who "believe that solutions emerge from your judicious study of discernible reality." I nodded and murmured something about enlightenment principles and empiricism. He cut me off. "That's not the way the world really works anymore," he continued. "We're an empire now, and when we act, we create our own reality. And while you're studying that reality — judiciously, as you will — we'll act again, creating other new realities, which you can study too, and that's how things

> will sort out. We're history's actors … and you, all of you, will be
> left to just study what we do."[3]

A state in which facts are replaced by a constructed reality formed by an empire's actions and narratives is rather extreme. But the post-factual democracy is indeed an extreme situation at the limit.

The concept of post-factual democracy is to be understood as one extreme (or a limit point) of a graded scale on which the purely factual democracy is a limit point on the other end of this teeter-totter.

6.3   Democratic Beacons

If a democracy at any given time and place is categorized as either factual or post-factual, you risk losing sense of the diverse social tendencies pulling in several directions all at once and creating a nuanced picture of what is real rather than a simple either/or situation. Societal development is not unambiguous. In order to navigate in a forever changing and messy reality and understand the tendencies and phenomena at play in our time, there is a need for maps with guideposts and beacons to navigate properly. With such beacons it will be possible to gain understanding of a complex and changing world that may form the foundation for further study of the political landscape. The concepts of factual and post-factual democracies are such beacons; they are *ideal types*. Sociologist Max Weber (1864–1920) introduced the ideal type as a conceptual instrument to compare different singular phenomena (Coser 1977). According to Weber, ideal types are methodological tools to analyze the world, not describe it in detail:

---

[3] Suskind, R. (2004): "Faith, Certainty and the Presidency of George W. Bush," *New York Times*, October 17, 2004. Verified June 11, 2017: http://www.nytimes.com/2004/10/17/magazine/faith-certainty-and-the-presidency-of-george-w-bush.html

"In its conceptual purity, the mental construct cannot be found anywhere empirically in reality. It is a utopia."[4]

Ideal types, among them factual and the post-factual democracy, are not realistic one-to-one mappings of political reality. They are beacons assisting in delineating tendencies and developments in a complex social reality. In a normative sense, both ideal types are more dystopian than utopian. Neither post-factual nor factual democracy is especially democratic.

Ideally, a democracy is based on a division of labor between uncovering facts much up to journalists, legal bodies, and scientists, and the democratic deliberation and debate carried out between politicians and citizens fully equipped with values and visions for the good life and the just society. In both post-factual and factual democracy, this division of labor is all but a lost tale.

6.4   Division of Labor as Ideal

There are political opinions about facts: opinions as to whether or not they are fair, whether or not they should be changed, and in what direction and what way they should be changed. But questions about whether facts are indeed facts are not political questions; they are determined by consulting science, law, or journalism. If factual matters are made political, the division of labor has broken down. Upholding the division of labor requires a certain amount of respect from politicians for the institutions and methods that reliably deliver knowledge. It is imperative not to discredit scientific results and researchers simply because they run counter to political interests and agendas.

The division of labor is not absolute: Knowledge about society is not disjoint from political discussions about what society ought to be. Both scientists and journalists have a limited amount of attention at their disposal. Attending to

---

[4] Weber (1949: p. 90).

one thing you are not attending to another in the zero-sum game of attention allocation. It is a choice of what is considered important—and that is not value-neutral. Pure "positive science" (Friedmann 1953) and value-free journalism are impossible. But objectivity and neutrality are ideals to aim at.

On the one hand, researchers and experts have a certain authority regarding facts. On the other, this does not imply that researchers and experts are always right. They make mistakes, research may be poorly executed, and research methods may be dissociated from facts and reality. In wake of the latest financial crisis, movements of students and researchers have come together, including well-known experts, critical of the hegemonic paradigm in economics. Nobel Prize winning economist Paul Krugman is accusing economics of having lost touch with reality mainly by promoting *the best of all possible worlds* where markets operate ideally and where mathematical beauty has been mistaken for truth.[5] The movement is working for reforms of theories, methods, and education in economics.[6]

There is a fundamental difference between critique that is coming from within a scientific field and the rejection of objectivity and expertise based on disbelief and distrust in researchers and experts. The latter might develop into an outright conspiracy theory vis-á-vis the rejection of the findings in climate research. Even though experts and institutions that produce knowledge are fallible, it doesn't mean that the ideal of a division of labor is outdated. But the fallibility calls for scientific humility and openness.

---

[5] Krugman, P. (2009). "How did economists get it so wrong?" *New York Times Magazine*, October 2nd 2009. Verified February 4th 2017: http://www.nytimes.com/2009/09/06/magazine/06Economic-t.html

[6] By way of example, *Institute of New Economic Thinking* (https://www.ineteconomics.org/) and its worldwide student network *Young Scholar's Initiative* (https://www.ineteconomics.org/community/young-scholars?p=community/young-scholars), the international movement *Rethinking Economics* (http://www.rethinkeconomics.org/) and *Evonomics* (http://evonomics.com/).

Division of labor requires scientists and experts not to turn political issues into mere matters of a scientific or technical, i.e., factual and kind. In the extreme case of an entirely fact-based democracy, there are no political issues and no room for differing, but legitimate, opinions. All issues are instead made to be a question of facts simply requiring a response from a scientific expert. Factual democracy is not very democratic either; it is a technocracy.

6.5   The Factual Democracy Is Technocratic

For Enlightenment philosopher Francis Bacon, it did not suffice that knowledge in itself is power and puts man in the position of mastering nature; those who possess knowledge must also rule politically. Bacon's utopia is described in *The New Atlantis*[7] from 1627, with its enthusiastic story about the invented country Bensalam. Even though a king is mentioned in the story, the country is run by a council of scientists, the Fathers of Salomon's House. Bacon's ideal state was ruled by scientists and experts and had no real political processes (Burris 1993); Bacon dreamt of technocracy.

In a technocracy, all issues are turned into questions of facts. If even normative, value-based matters related to how society ought to be are turned into factual matters for science and experts to decide upon, there is nothing to debate democratically and nothing to have a political opinion about. Citizens have but to follow the experts' directions. If they do not, they not only disagree, they are wrong.

The European Union (EU) demonstrated technocratic tendencies as to the harsh austerity policies that it, and especially Germany, used as forced means of addressing the debt crisis in troubled countries such as Italy and Greece. These countries were furnished technocratic governments to implement the austerity deemed necessary. Turning political deci-

---

[7] Accesible at *Projekt Gutenberg*, verified June 14, 2017: https://www.gutenberg.org/files/2434/2434-h/2434-h.htm

sions into technical directives based on membership in the eurozone, making them exempt from political debate, undermines democracy, not least when the resulting policy has considerable economic and social consequences for those affected, primarily low-income groups of people:

> Technocrates can be very apt at saying how much [economic] pain a country may endure, how the debt level may be made endurable, or how to solve a financial crisis. But they are not good at finding out how to spread the pain, whether to increase the taxes, or if it is necessary to cut down costs for one group or another, and what the consequences of the chosen policy are on the distribution of income. These are political questions, not technocratic.[8]

Public anger may arise from the tendency for years on end to employ somewhat too factual and technocratic policies that lack the sense and acknowledgment of the pain they cause to publics. Post-factual tendencies and symptoms may be partially motivated by anger, and the anger might be for a reason. "Britain has had enough experts" was the slogan with which Michael Gove, UK Environment Secretary and Brexit supporter, phrased the general distrust of the political system up to the Brexit vote. As German philosopher Jürgen Habermas said in 2013, the EU is caught between "the economic policies necessary to keep the Euro on the one hand, and the political steps towards a closer integration on the other. This means that necessary steps create resentment and meet spontaneous popular resistance."[9]

Even though post-factual tendencies have succeeded too factual democracy, it does not mean that the factual democracy deserves our nostalgic longing for it. The factual democracy is not some democratic Golden Age. If post-factual symptoms and situations become more permanent, however,

---

[8]"Have PhD, will govern," editorial, *The Economist*, November 16, 2011. Verified June 11, 2017: http://www.economist.com/blogs/newsbook/2011/11/technocrats-and-democracy

[9]Traynor, I. (2013): "Habermas advarer: Tyskland sætter Europas liberale demokrati på spil," *Information*, April 30, 2013. Verified June 11, 2017: https://www.information.dk/udland/2013/04/habermas-advarer-tyskland-saetter-europas-liberale-demokrati-paa-spil

a decay of democracy may occur, where the powers that be are not accountable even if they are caught lying through their teeth.

6.6 Lies and Deceit

It is not breaking news that politicians twist or conceal the truth, playact, deceive the public, talk bullshit, and lie. Those are standard elements in the political game. Also, no Golden Age ever existed in which the politicians were all honest, authentic, and always truthful. Nevertheless, being caught lying or deceiving used to be something to avoid at all costs. The father of modern political theory, Niccolo Machiavelli (1469–1525), harbored no rose-tinted illusions regarding politics and the brutal game in the fight for power. All is fair in raw power politics. Machiavelli's protagonist, the Prince, in his book entitled the same, needs to be able to act both as lion and fox, to show the raw power, brutality, and strength of the former but also be sly and avoid traps like the latter. Deceit is necessary to obtain and hang on to power (Machiavelli 1999). However, it is important not to get caught; one's lies must resemble truths. The Prince must therefore hone his skills at playacting, deceit, and hypocrisy. Machiavelli thus instructed a politically ambitious diplomat in a correspondence: "Occasionally words must serve to veil the facts. But let this happen in such a way that no one become aware of it; or, if it should be noticed, excuses must be at hand to be produced immediately."[10]

This is usually a very good advice. Getting caught lying or being untruthful has traditionally cost politicians their careers or at least cost them *something*. But, caught being untruthful is not very damaging if your voters do not see, read, or believe

---

[10] Machiavelli, N. (1882): *The Historical, Political, and Diplomatic Writings of Niccolo Machiavelli*, tr. from the Italian, by Christian E. Detmold. Vol. 4. Boston: J. R. Osgood and company. Pp. 422. Verified February 5, 2017: http://oll.libertyfund.org/titles/777

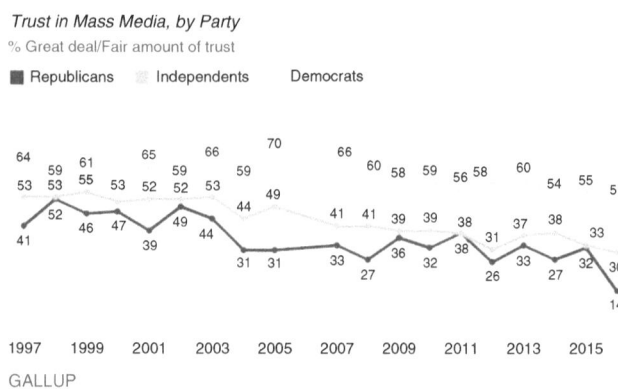

F<small>IG</small>. 6.2. The amount of trust that American citizens have in the mass media, divided by partisanship. (Swift, A. (2016): "Americans' Trust in Mass Media Sinks to New Low," *Gallup*, October 14, 2016. Verified June 11, 2017: http://www.gallup.com/poll/195542/americans-trust-mass-media-sinks-new-low.aspx).

the media or fact checkers that reveal it. Or if you have sufficiently strong loyalty from your constituency, and the polarization is deep enough for the media to be seen as nothing short of the enemy to whom it is only fair to tell blue lies. Or if the news media are declared to be fake news when their coverage does not suit you, and your constituency has so little trust in the media that they accept your claim (Fig. 6.2).

Facts become secondary to political success if enough people do not trust what is reported to be facts. When the distrust reaches a certain threshold, the result is skepticism that undermines the fact-based evaluation of the politicians in power and the capacity to hold them accountable accordingly. *If everything is a lie anyway, then one liar is not worse than the other, and I prefer my liar to yours.* Distrust at this level undermines democracy.

6.7 Accountability

In a democracy, the people rule. The term itself is witness. A necessary condition for democracy is that representatives of the people, the politicians in power, are accountable to the people. If the citizens cannot at the very least hold the politicians accountable by firing them, then the people do not rule, and it is not a democracy. In a minimalist model of democracy, the population's political preferences may be seen as input, and the chosen policies and legislation as output of the democratic system (Fig. 6.3).

When politicians do not rule according to the population's political preferences, and if they are not responsive to them in their policies, they are poor representatives for the population in question. In that case, they may be held accountable and might even be replaced by the voters. Election day is the day of reckoning.

Alongside the public's possibility to hold the politicians in power accountable on election day, the politicians in power are subjected to an institutionalized checks and balances across the bodies of governmental power and electoral periods. The Danish invention of the Ombudsman has been an export success, and this is such an institution. It was established to keep an eye on the politicians' actions pertaining to best practice and legal administration practice (Kriesi et al. 2013). Control mechanisms vary from country to country and democracy to democracy. In the USA, the principle of the division of power takes the form of three branches—legislative, executive, and judicial—with an institutionalized system of checks and balances.

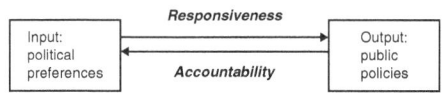

FIG. 6.3. A minimalist model for representative democracy. (Source: Kriesi et al. 2013).

In addition to the judicial power's ability to overrule laws, the legislative power (in the USA, the Congress) may seek to impeach the President if there is a suspicion that the President is guilty of "high crimes and misdemeanors," the standard articulated in the US Constitution. However, holding a president legally accountable in this manner, except in extreme cases like the Watergate scandal that led to President Richard Nixon's resignation (when it was clear that even the Senate Republicans would not vote to prevent his conviction), may well depend upon the political makeup in Congress. Generally, then, if there is no majority in the legislature to hold the executive power accountable, offenses may receive no consequence, even if discovered. And checks and balances only work if the political majority is more loyal to the law or the Constitution than to the party, the secretary, minister, the government, or the President. If these checks don't work, the voters may settle the score on election day and ensure a new majority. But whether voters actually hold the politicians in power accountable for what they have done, failed to do, and have promised to do depends on:

1. Factual information about what they have done or not done being circulated sufficiently to reach the voters
2. The voters having, for good reason, sufficient trust in the media that bring the information
3. The voters acting accordingly on election day

If the democratic institutions making it possible to hold the politicians in power responsible based on facts (i.e., the media, Congress, and the courts) are undermined, that in turn undermines democracy. Undermining the legitimacy of the media and the courts is to undermine the trust in the control mechanisms put in place to avoid democracy devolving into an authoritarian regime. If the politicians try to fire the watchdogs and the gatekeepers in order to stay in office and not be held accountable, that is the equivalent of breaking democracy's basic contract.

6.8   Totalitarian Propaganda

On May 26, 2017, President Trump sent an e-mail to all who had registered for his mailing list. The subject was "Stop the FAKE NEWS."

Drafts ⌄
Dato: 26. maj 2017 8.09 PM
Emne: Stop the FAKE NEWS
Til: *carolinehsaabye@gmail.com* <carolinehsaabye@gmail.com>
Cc:

Friend,

I've said it before and I will say it again: **the Fake News Media is the real opposition**.

It's a 24/7 barrage of hit jobs, fake stories, and absolute hatred for everything we stand for as a movement.

And the same talking heads that said Hillary Clinton had a 99% chance of winning the election now think they can speak on behalf of the American people. It's time to once again release our Mainstream Media Accountability Survey to show the that the American people are fed up with the Fake News Machine.

I need you to take the MEDIA ACCOUNTABILITY SURVEY to do your part to fight back against the fake news attacks and deceptions.

They don't care about the truth. They don't care about what's right. They only care about propping up the liberal Democrats they worship.

There is nothing they won't do to stop us.

This is a fight we can't afford to lose. The future of America hangs in the balance. Our country is at stake.

Please take the Mainstream Media Accountability Survey to do your part to fight back against the media's attacks and deceptions.

It is already subversive for democracy as a President declares the media, holding politicians and leaders accountable to the public, a sworn enemy, and makes that very same public hold the media accountable instead. When you add that the supporters are labelled "a movement" and presented as central for an existential and epic struggle for America's future and survival, with the media on the side of the enemy, the rhetorics begin to show traits of totalitarian propaganda.

According to Hannah Arendt (1906–1975), the propaganda of a totalitarian movement striving for power employs simple and coherent identity forming and meaning creating narratives. Such narratives offer the otherwise stigmatized and alienated supporter of the totalitarian movement a role to play and have a purpose in a thus existentially meaningful, albeit fictitious, "alternative" pseudo-reality. Totalitarian propaganda feeds on *us*-versus-*them* narratives and utilizes distrust, tribalism, polarization, and conspiracy theories as weapons in the struggle for power. Narrative coherence, a sense of purpose and meaningfulness, belonging to a group and playing a role in the struggle between good and evil, may in extremis make us ignore even what our own senses tell us. And, according to Arendt, the propaganda serves exactly that purpose:

> The propaganda of the totalitarian movement also serves **to emancipate thought from experience and reality**; it always tries to inject a secret meeting in every public, tangible event and to suspect a secret intent behind every public political act. Once the movements have come to power, they proceed to change reality in accordance with their ideological claims. The concept of enmity is replaced by that of conspiracy …[11]

Creating a media and information environment of distrust and conspiratorial suspicion may make way for immunizing oneself to legitimate critique and avoid being held democratically accountable. When the public's trust in the sources providing reliable information is sufficient low, an authoritarian figure can define what is real and make up the facts suited for gaining necessary support for seizing and consolidating power. If facts and evidence have lost all authority, as a limiting post-factual state, it may contribute in giving way for a rule where self-determination is replaced by loyalty to the leader and identification with the movement:

> The ideal subject of totalitarian rule is not rule is not the *convinced Nazi* or the *convinced Communist*, but people for whom the distinction between fact and fiction (i.e., the reality of experi-

---

[11] Arendt 1951: p. 585. Our emphasis.

ence) and the distinction between *true* and *false* (i.e., the stan-
dards of thought) no longer exist.[12]

To be epistemically emancipated from reality may be a step
toward the opposite of emancipation in a political context,
toward dominion and oppression. Post-factuality may be a
prelude to tyranny. To set oneself free from the real world is
a step toward being more easily controlled. This stems not
only from the new opportunities of creating and spreading
mis- and disinformation and creates distrust which digitalization of the media and information have made possible. A
factual society may be an even bigger threat to freedom and
autonomy than post-factual relativism and distrust. The
dream of digital emancipation may turn into a nightmare of
digital totalitarianism.

---

[12] Arendt (1951: p. 591).

.

Chapter 7
Epilogue: Digital Roads to Totalitarianism

7.1 Digital Emancipation

The digital revolution was meant to emancipate. In the *Declaration of the Independence of Cyberspace* from 1996, John Berry Barlow declares the new digital reality, Cyberspace, to be an independent new world of freedom and equality without oppression of the old world of nation-states ruled by governments. Barlow compares the digital revolution to the American War of Independence and the pioneers of digitalization to the heroes of the American Revolution: "… those previous lovers of freedom and self-determination who had to reject the authorities of distant, uninformed powers."[1]

The digital reality that came to pass the following years and we are now in the midst of is far away from the digital utopia of liberation and self-determination. The Internet and the digital technology may just as well pave ways for new forms of oppression and dominion. Instead of being a force of democratization and emancipation, the digital revolution may turn out being the opposite and contribute to undermine democracy and political self-determination.

---

[1] John P. Barlow (1996): A declaration of the independence of cyberspace. Verified January 7, 2018 https://www.eff.org/cyberspace-independence

© The Author(s) 2019   119
V. F. Hendricks, M. Vestergaard, *Reality Lost*,
https://doi.org/10.1007/978-3-030-00813-0_7

The Arab Spring in 2010 is an example of an event where the Internet and social media played an empowering and emancipatory role for citizens by providing a communicative infrastructure for the uprisings the authorities did not control. In Tunisia Facebook took hold as the "revolution headquarters" and in Egypt it served as online incubator of a revolutionary youth movement that could mobilize and organize protests (Herrera 2015). Four years after the uprising, tables have turned. In Egypt the military regime began in 2014 the so-called Social Networks Security Hazard Monitoring System operation, which is a surveillance program informing the regime of the whereabouts and communication of citizens much more efficiently than during the former authoritarian regime: a tool for emancipation and protest is turned into a tool of repression and social control.

It is not only democracy and self-determination on the political level that run the risk of being disrupted by the digitalization. In the end, the digital revolution may result in an elimination of autonomy and free will of the individual. Too much information may be a greater threat to freedom than misinformation. A digitalized too factual and too informed society may amount to a new form of digital totalitarianism. The lack of trust that fuel post-factual tendencies may be the least of problems compared to a data-driven factual society where trust is eliminated as phenomenon and replaced by control. It is on its way in China.

7.2   The Chinese Panopticon

In 2014 the State Council in China announced and initiated the construction of the *Social Credit System*, an ambitious project with the official purpose of generating "honest mentality," fostering a "culture of sincerity," and raising the "level of trustworthiness in the entire society." This is an important step toward building a "harmonious socialist society" and stimulating "the progress of civilization" in words of the official document.[2] The system is under construction, the details are not yet in place, and participation will first be mandatory for every citizen and company in China from 2020 onward. Nevertheless, test versions already running locally and statements from those responsible for the implementation draw a picture of what is in the making: a surveillance society made possible by the digital revolution that enables monitoring and disciplining of the citizens by constantly providing incentives to conform to desired behavior.

The baseline is a rating system. Every citizen and company will have an account for social credit points and the score will determine the trustworthiness of the particular individual or company.[3] One's credit score will be decided automatically employing artificial intelligence for analyzing an enormous amount of data and information about the citizens. The information is gathered from many different sources of data like banks and financial institutions, stores, public transportation systems, Internet platforms, social media, and e-mail accounts. Not to forget the 570 million surveillance cameras with integrated facial recognition that are to be in place in 2020 with 170 million already up and running today (Fig. 7.1).[4]

Digitalization and the Internet have enabled such massive data collection that surveillance may be almost total with no angles out of sight or blind spots: an updated digitalized ver-

---

[2] Planning Outline for the Construction of a Social Credit System (2014-2020). Posted on June 14, 2014 Updated on April 25, 2015 State Council Notice concerning Issuance of the Planning Outline for the Construction of a Social Credit System (2014–2020), GF No. (2014)21. Verified 24.06.2017: https://chinacopyrightandmedia.wordpress.com/2014/06/14/planning-outline-for-the-construction-of-a-social-credit-system-2014-2020/

[3] Kai Strittmatter (2017): Creating the honest man. München, Süddeutsche Zeitung. Verified January 7th 2018 http://international.sueddeutsche.de/post/161355280290/creating-the-honest-man

[4] Chan, T.F. (2018). "Parts of China are using facial recognition technology that can scan the country's entire population in a second," *Business Insider*, 27.05.2018, verified 26.06.2018: http://nordic.businessinsider.com/china-facial-recognition-technology-works-in-one-second-2018-3?r=US&IR=T

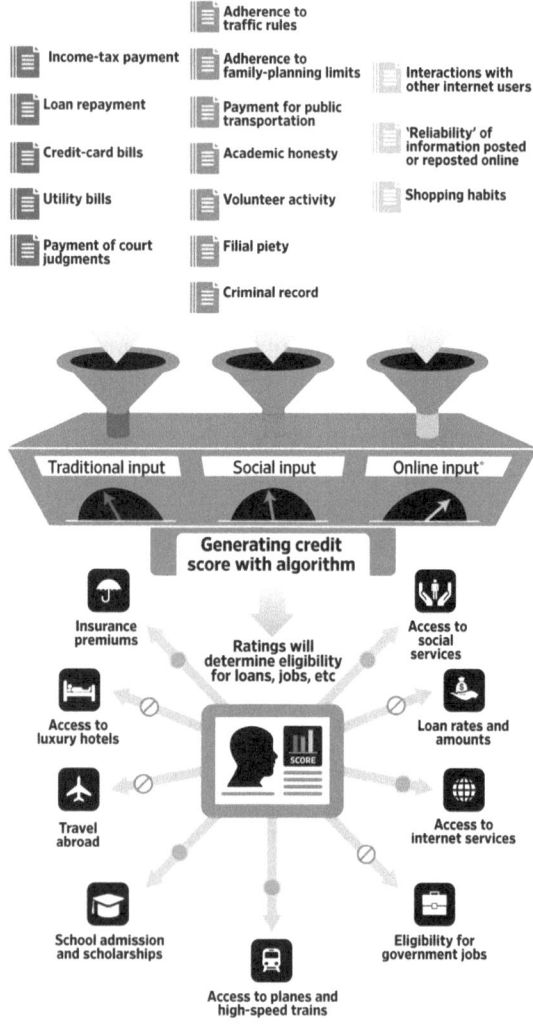

F<small>IG</small>. 7.1. Outline of the Social Credit System in China. (Source: *The Wall Street Journal*, 2016).

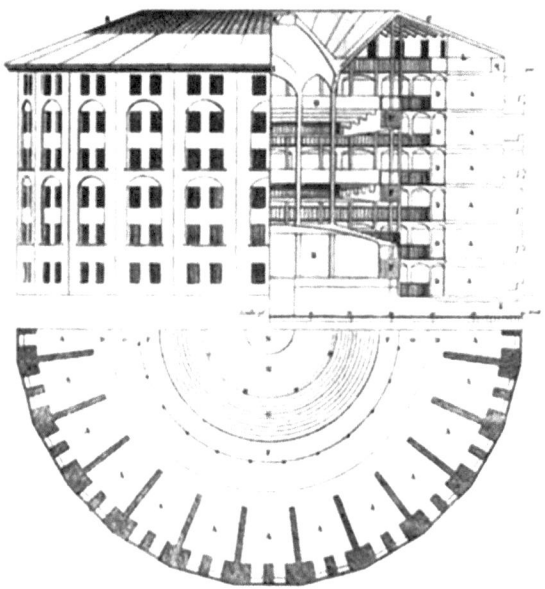

FIG. 7.2. Jeremy Bentham's original architectonical plan for a *Panopticon*. A prison in which inmates are monitored constantly with no blank spots in the cells to hide from the Guards' gaze. (Wikipedia Commons. Verified 24.06.2018: https://commons.wikimedia.org/wiki/File:Penetentiary_Panopticon_Plan.jpg).

sion 2.0 of Jeremy Bentham's *Panopticon* ("all-seeing") where one's life is being monitored in detail 24-7 without any possible refuge from the gaze of the authorities. In the Panopticon, privacy is not an option (Fig. 7.2).

The Panopticon is employing surveillance to discipline the inmates to desired behavior. In Bentham's own words:

The more strictly we are watched, the better we behave.[5]

_____

[5] Jeremy Benthem: unpublished, from the manuscripts of Jeremy Bentham in the Library of University College London. Leaflets on Bentham's Life and Work. Verified 24.06.2018: https://www.ucl.ac.uk/bentham-project/publications/leaflets-benthams-life-and-work

Surveillance contributes to discipline the surveilled (Haidt 2012). With surveillance and the information that results from it comes a disciplinary form of power (Foucault 1979). Monitoring and registering combined with rewards and sanctions for wanted and unwanted behavior is an efficient tool of *behavior modification*. Whether the surveillance takes place in institutions as prisons, schools, hospitals, the workplace, or the army or is generalized to the whole society, it aims at the same result: the production of obedient subjects who conform to what is expected and wanted by the authorities—normalized and law- and norm-abiding citizens.

The Social Credit System employs this panoptic power technique of surveillance-based discipline. The system monitors, registers, and rewards desired behavior with adding social credit points to one's credit score. Unwanted behavior, on the other hand, will cost points and lower the score. Voluntary community service, taking care of one's family, charity donations, and responsible consuming like buying diapers are point rewarding. On the other hand, if you, for instance, spend too much time and money on computer games, smoke in a nonsmoking zone, drop a cigarette bud, travel without ticket, drive recklessly, miss paying a bill in time, or spread fake news on social media, you lose points and the score drops. In one running version of a Social Credit System, Sesame Credit, an additional feature is that online friends on social media also count. Their score reflects on one's own providing an incentive to restrict one's relations to only model citizens deemed trustworthy. According to the 2014 plan, "reporting" others' breach of trust—as it is phrased—will also be rewarded.

Besides being publicly assessable for other citizens and their evaluation of one's character, the social credit score will determine one's economic and social opportunities and restrictions. It is decisive for whether one, for instance, may obtain a loan, get a job, enroll one's kids in a good school, or have access to public services. Social sanctions may also apply. Citizens with a low score risk being publicly exposed and shamed on billboards and social network sites as morally flawed people, whereas those with a high score are presented and promoted as model citizens making it easier to find a partner. The score also decides one's freedom of movement. Restrictions that approximately 9 million Chinese with a low score experienced the hard way when they were denied booking tickets for domestic flights and high-speed trains. As the official catchphrase bluntly states:

> Allow the trustworthy to roam everywhere under heaven while making it hard for the discredited to take a single step.[6]

It seems to work. Where the test versions are in operation locally, it has been reported that people's behavior and the social environment have changed for the better. The change of behavior results in the beginning from a conscious calculation and cost-benefit analysis: it pays off to behave well. After a while, though, the rules, regulations, and norms are internalized into an unconscious habit. As a citizen puts it: "At first, we just worried about losing points, but now we got used to it."[7]

Some even have the ambition and hope that it will work so well it is able to change human nature itself and create a new more honest and better human being and citizen. Zhao Ryuing, who is in charge of the implementation in Shanghai, envisions that the system eventually will eliminate not only the need for punishment but also a social *thoughts*:

> We may reach the point where no one would even dare to think of breaching trust, a point where no one would even consider hurting the community. If we reached this point, our work would be done.[8]

If this goal is ever achieved, there will no doubt be social order and maybe even "harmony." The price to pay, on the

---

[6] Kai Strittmatter (2017): Creating the honest man. München, *Süddeutsche Zeitung*. Verified January 7th 2018 http://international.sueddeutsche.de/post/161355280290/creating-the-honest-man

[7] Mistreanu, S. (2018). Life Inside China's Social Credit Laboratory, *Foreign Policy*, 18.03.2018. Verified 24.06.2018: http://foreignpolicy.com/2018/04/03/life-inside-chinas-social-credit-laboratory/

[8] Ibid. (Stritmatter 2017)

other hand, is the actualization of a totalitarian Big Brother state that monitors, registers, and reprograms its citizens' behavior to total obedience with no space left for even unwanted thinking. Such a result, where even *thought crimes* are eradicated, would make even the leading members of The Party in the novel *1984* a little bit jealous.

Big Brother's methods of control have been updated. Even if China is not soft on crime, the Social Credit System employs power techniques of rewards and desire for happiness and success rather than fear, terror, and violence of old-school totalitarianism. Totalitarianism with a human face resembles *Brave New World* more than the brutality of *1984*. To condition and motivate may be a much more efficient way to produce obedient and conforming citizens than to repress through fear and terror. The new methods of sugar-coated digital control may show so effective they succeed in undermining critical thinking, autonomy, and self-determination. Not just in China.

The Chinese State Council is not alone in totalitarian dreams of total surveillance and control. It does not necessarily require an authoritarian state aiming at social order and harmony as primus motor to reach totalitarian conditions. An unregulated market for data and user information and the hunt for profits may be sufficient. Google is leading the way in a race and mission of colonizing, commodifying, and monetizing every aspect of our life.

7.3   Surveillance Capitalism

The same year China initiated the construction of the Social Credit System, 2014; on the other side of the Pacific Ocean in Silicon Valley, Hal Varian, chief economist at Google, held a speech addressing the great opportunities made possible by extraction of data combined with massive processing power. The speech revealed a vision for a future of a surveillance capitalistic society with no more room for privacy than in China. According to Varian, the elimination of privacy is the (fair) price to pay, not for social order and harmony, but for the functionality, efficiency, and convenience of the products and services Google provides to the users. Gathering and analyzing information *about* the user is the condition for personalizing the products *for* the user. Privacy is exchanged for enhanced user experience and the convenience made possible by personalized optimization and customization of the products that automatically tailor them to the individual user. Only when Google knows at least your location, budget, and food preferences, it is able to provide a relevant suggestion for a restaurant nearby to have dinner. The more personal information Google possesses, the more efficiently and conveniently it is able to serve one's individual needs and demands. That is the basic justification of the surveillance. However, even more surveillance, data mining, and information gathering are necessary to realize the tech giant's vision. The ambition is that Google products should run so smoothly and conveniently for the user that it is not even necessary to *google* or ask Google questions. As company founder, Larry Page is quoted for saying:

> [Google] should know what you want and tell it to you before you ask the question.[9]

Google should be able to predict our wants and desires before we have articulated them. To be able to do that, it needs to know us almost better than we know ourselves. That takes *a lot* of information. It also implies a total elimination of privacy. That is the necessary condition for mining the amount of data sufficient for knowing, predicting, and attending to our wants and desires before we have expressed them ourselves. For a company like Google, the right to privacy is an obstacle to their ambition and stated mission. To "organize the world's information and make it universally accessible and useful"[10] taken literally implies a colonization and dispos-

---

[9] Hal R. Varian (2014): *Beyond Big Data.* Business Economics 49 (1)
[10] Google.com/ about our company. Verified 24.06.2018: https://www.google.com/about/our-company/

session of every bit of our private sphere and life: accessibility excludes privacy. For the right to privacy, Google CEO Eric Schmidt's defense for the surveillance practices is telling and chilling:

> If you have something that you don't want anyone to know, maybe you shouldn't be doing it in the first place.[11]

Google does not have an extended network of surveillance cameras to gather data and de facto implode the difference between being online and offline into an *Onlife*, as in the Chinese surveillance society. Nor is it mandatory to use Google products. Nevertheless, the vast amounts of data necessary will increasingly be accessible through *the Internet of Things*. With "smart" products like "intelligent" clothes, household appliances, fitness equipment, toys, personal assistants, learning devices, etc., our whole lifeworld becomes more and more embedded in a fine-grained network of sensors able to monitor and register everything we say or do.[12] For the sake of shiny things as convenience, optimization, and "enhanced user experiences," we are piece by piece trading away the former virgin territory of our intimate social life. Device by device, we are building our own privately run, commercial Panopticon with no space left for privacy. Neither will it be possible to *opt out* even if it is not mandatory and no one forces one to buy anything. Already today, the Internet is deeply embedded in our societies and our social, communicative, economic, and politically infrastructure. Getting an education, finding employment and securing an income, having a credit card and a bank account, participating politically, ad communicating and interacting socially without being a part of the digital infrastructure is uphill to say at least. With the development of the Internet of Things and "smart cities," going into the wilds may be the only alternative to total surveillance—which is no real alternative at all.

---

[11] Google CEO On Privacy (VIDEO): 'If You Have Something You Don't Want Anyone To Know, Maybe You Shouldn't Be Doing It', *Huffington Post*, 18.0.2018. Verified 24.06.2018: https://www.huffington-post.com/2009/12/07/google-ceo-on-privacy-if_n_383105.html

[12] See Chap. 1, Sect. 1.7.

Google is one of many. The business model based on data mining, Google pioneered, has become a standard business model and yardstick for new startup companies, providing a service, some entertainment, a utility, or information in exchange for user data. There is a data fever going on comparable to gold rushes of the nineteenth century.[13] With a lot of different actors racing for getting a piece of the new action—and as the data assets are mined and appropriated at this new frontier of commercialism and commodification—we are dispossessed of the information our behavior produces. We are alienated from the value our data creates and lose control. Data enables prediction and prediction makes for control. The data we are providing is used to control us.

7.4   Prediction is Profit

The main reason data is valuable when pooled and aggregated into Big Data is that it enables predictions of the future. Data provides patterns of past behavior that may show probable future behavior. It makes possible data-informed calculations of risk, future sales, gains and expenses, effects of marketing, and communication strategies as well as on how to optimize communication, marketing, and design. If you are able to calculate probable future behavior, you are able to make a profit from it:

> Will she be able to repay the loan in the future – and will she? Will he show up for work and contribute to productivity or is he just a "high cost" employee? Is she disposed for a disease, so the insurance company in case a policy is made has to cover medical expenses exceeding the income from premiums? Which commercials will succeed persuading this person to buy the product or vote for the candidate? How many more users will push the button and provide valuable attention, if it is red? Which products and services will he desire later today? In two minutes? Ten seconds after this specific online marketing stimulus is provided through his smartphone?

---

[13] Jankowski, S. (2014). The Sectors Where the Internet of Things Really Matters, *Harvard Business Review*, 22.09.2014. Verified 24.06.2018: https://hbr.org/2014/10/the-sectors-where-the-internet-of-things-really-matters

All these questions come down to the same: how may future profits be generated and maximized? Data-driven prediction of behavior for the sake of sales and profit is the essence of the leading surveillance capitalism business model.[14] Fortune telling has become big business. Prediction is profit.

Profitable predictions may become chains restraining and undermining the self-determination of the citizen. In a free data market unrestrained by effective citizen protection laws, the traditional *financial* credit rating scores may start to look like the all-embracing *social* credit scores in China.

With Big Data extraction and analyses, the market for consumer and credit information in the USA has become the Wild West. The traditional credit bureaus formerly operating in regulated territory have transformed into data brokers with the will and ability to circumvent the law—at the expense of citizens' rights. Whereas individual level information is regulated by law, the information it is possible to extract through data mining is not. With sufficient data, it's possible to calculate information just as sensitive as individual level information. An event title in 2011 by two of former American credit bureaus turned data brokers, FICO and Equifax, is telling: "Enhancing Your Marketing Effectiveness and Decisions With Non-Regulated Data."[15] In the new unregulated domain of data-driven credit scores, the approach is that "all data is credit data."[16] With such an *anything-goes* approach, profile data, and online social footprints, the device you use and how quickly you scroll through the sites are nowadays factors that may feed into your credit score. Facebook has taken it further

---

[14] Zuboff, S. (2016). "The Secrets of Surveillance Capitalism", *FAZ*, 05.03.2016. Verified 24.06.2018: http://www.faz.net/aktuell/feuilleton/debatten/the-digital-debate/shoshana-zuboff-secrets-of-surveillance-capitalism-14103616.html?printPagedArticle=true#pageIndex_0

[15] Taylor, A. & Sadowski, J. (2015). "How Companies Turn Your Facebook Activity Into a Credit Score", *The Nation*, 27.05.2017. Verified 24.06.2018: https://www.thenation.com/article/how-companies-turn-your-facebook-activity-credit-score/

[16] Lauer, J. (2017). *Creditworthy: A History of Consumer Surveillance and Financial Identity in America*. New York: Columbia University Press: 267.

and has patented a method to calculate credit scores based on one's social network, so the average credit score of one's friends is decisive for one's own (Hurley and Adebayo 2017). In a capitalist country as the USA, access to credit is a make-or-break for everyone but those in the top 1% of the wealth and income scale. To buy a house, you need a mortgage; you need a loan for buying a car and getting a college degree for yourself or your children depend on access to credit and student loans. Employers also consider credit rating scores when hiring and so do property owners evaluating potential tenants. With no effective legal restrictions on gathering and usage of data, the door is open for insurance companies gathering data on health and habits undermining the whole idea of pooling risk in insurance products and making it impossible to get insured for the ones most in need. Your credit rating score is decisive for the opportunities you have and the restrictions you face. If there is no limit to what goes into one's credit score and how it is used by creditors and banks, landlords and car rental companies, and employers and insurance companies, the difference between financial credit rating and *social* credit rating in China is diminishing.

7.5   Prediction Is Power

Prediction is power. If you are able to predict the future, you may also be able to influence and change it—and make a buck doing so. Predict to change is the core of targeted marketing. Marketing success is success in changing people's behavior in a commercially or politically profitable way for the client. To be able to predict behavior makes it possible to change and modify it by providing the right stimuli at the right moment. If you can predict what people want and when they want it, it is possible to precision nudge and steer them to buy it from you. The more you know, the better you are able to predict, and the better you predict, the more successfully you may influence and control. Already demographic profiling grants this kind of power. Predictions made from data such as home address, gender, ethnicity, employment,

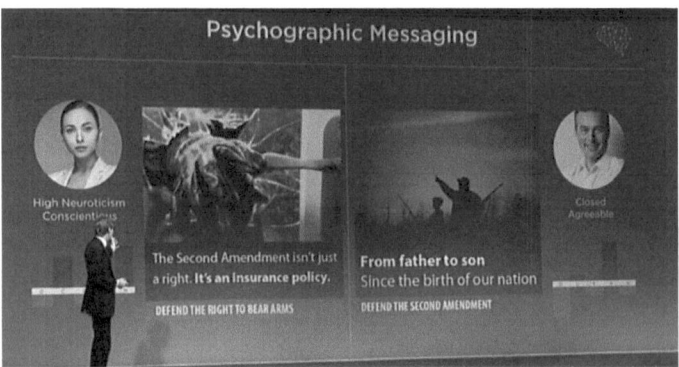

FIG. 7.3. Former CEO of *Cambridge Analytica*, Alexander Nix, illustrates the potential in psychological profiling. The right to carry arms is the message to be sold, and if you are the fearful type (high score on neuroticism), the ad will play on fear of burglary and the right to carry arms is framed as an "insurance policy" (left). If, however, you are profiled as "closed" or tradition-bound, but "good-natured," the political marketing is tailored to *this* profile and the right to carry arms is framed with a hunting metaphor, patriotism, and family values: "From father to son. Since our nation was born" (right). (Wozniak, K. (2017): "Did Big Data Win the Election for Trump?", *Misciwriters*, April 18, 2017. Verified June 14, 2017: https://misciwriters.com/2017/04/18/did-big-data-win-the-election-for-trump/).

income, consumption patterns, political affiliations, and social network of family and friends make it possible to target and tailor ads to hit the pain points where it hurts. An effective predatory method for influencing people's behavior.[17] However, when the profiling moves under the skin and becomes psychological profiles of people's mental makeup and emotional life, it gets even more powerful and potentially oppressive.

Before the scandal, Cambridge Analytica also boasted that they are taking things a step further with profiling, integrating methods and results from scientific psychological research to

---

[17] See Chap. 1, Sect. 1.8.

create *psychological* profiles of users/consumers/voters/citizens. If you can categorize people according to their personality type and mental makeup, the targeted marketing bombardment may be conducted with even more precision and effect. Psychological profiling opens up chilling possibilities of affective management and emotional control. By employing, for instance, fear-mongering messages for someone profiled as a fearful personality type, it is possible to hit the target's pain points where it *really* hurts (Fig. 7.3).

Even if Cambridge Analytica was the one scandalized company that got caught, it is not alone with ambitions of affective influence and emotional control made possible by psychological profiling. It is highly potent, according to Alexander Polonsky, from the French data broker company Bloom:

> You can do things that you would not have dreamt of before. It goes beyond sharing information. It's sharing the thinking and the feeling behind this information, and that's extremely powerful.[18]

Powerful for *whom*, one might ask. Not the user, costumer, or citizen who is psychologically profiled to be effectively influenced and manipulated. The dream of a data broker may turn to a nightmare for the citizens. The scare of *The Hidden Persuaders* (Packard 1960) steering us secretly through subliminal influences in the 1960s may turn out to be fully justified in the age of Big Data and psychographics. Psychological profiling takes the knowledge that is power to the next level. Most of our life and behavior is governed by fast, automatic, involuntary, and unconscious mental processes evading our attention and awareness (Kahneman 2011). Humans are affective beings, rather than rational agents, and more controlled by emotions than we are in control of them (Haidt 2012). If you are able to influence those processes and associations, affects, and emotions going on in the dark basement

---

[18] Confessore, N. & Hakim, D. (2017). "Data Firm Says 'Secret Sauce' Aided Trump; Many Scoff," *New York Times*, 06.03.2017. Verified 07.01.2018: https://www.nytimes.com/2017/03/06/us/politics/cambridge-analytica.html

of our *psyche*, you can more or less control us. If one is taking advantage of people's deepest fear, makes them angry, or otherwise plays emotionally on fundamental personality traits, we may not even be aware of, it may undermine individual self-determination, rational agency, and autonomy. If companies—or the state for that matter—know us better than we know ourselves, an "emotional dictatorship" governing us without our knowledge or consent is right down the road. As conceived by the South Korean-born German philosopher and writer Byung-Chul Han:

> "Big Data" enables prediction of human responses and the future, therefore, can be manipulated accordingly. Big Data has the ability to turn people into puppets. Big Data generates knowledge that enables ruling power. And it is Big Data that makes it possible to access and manipulate the human psyche without the affected person being aware of it. Big Data essentially spells the end of free will.[19]

If this extreme situation of a data-driven total elimination of free will, individual sovereignty, and autonomy is ever fully reached, it would be the opposite of emancipation. Total predictability makes for total control. Freedom from uncertainty is not freedom at all. On the contrary, it is the stuff totalitarianism is made of.

7.6   Roads to Totalitarianism

According to Hannah Arendt, the aim of totalitarianism is total, unlimited power. This kind of power demands that everybody is "dominated in every aspect of their life" (Arendt 1951: 456). The greatest obstacle to that ambition is the unpredictability of humans due to our spontaneity, creativity, and freedom. Those must be eliminated to produce predicable subjects and humans reduced to bundles of conditioned reflexes controllable by stimuli to provide the desired

---

[19] Han, B.-C. (2016). "Digital Totalitarianism: How Big Data Is Killing Free Will", WorldCrunch, 07.05.2016, verified 26.06.2018: https://www.worldcrunch.com/culture-society/digital-totalitarianism-how-big-data-is-killing-free-will

and predicted response. Thus, the aim of totalitarianism is identical to a real-world actualization of behaviorism's deterministic understanding of human beings: stimuli-response. With the pace and acceleration, new digital technology is developed and integrated with behavioral science and design, and we may be heading toward—but hopefully never reach— a digital totalitarianism: *the total elimination of autonomy and self-determination by data-driven behavioral control.*

Long before the digital revolution, Arendt worried that the historical tendency is not leading us toward emancipation and a realization of freedom, but to the opposite:

> The trouble with modern theories of behaviorism is not that they are wrong but that they could become true, that they actually are the best possible conceptualization of certain obvious trends in modern society. It is quite conceivable that the modern age — which began with such an unprecedented and promising outburst of human activity — may end in the deadliest, most sterile passivity history has ever known. (Arendt 1951: 345)

With the brave new digital world and the marketization of user data, this worry has not become less pertinent.

Traditionally, totalitarianism is identified with *state* totalitarianism and characterized by the abolishment of any distinction between the state and civil society. Nothing outside and exempted the total dominion of the state.[20] The "total state," according to its advocate, Carl Schmidt, "embraces every domain" with the result that everything is "potentially political" (Schmidt 1932: 22). It is *the total interpenetration of the political and social and of state and society.*

It does not take a *state* with totalitarian ambitions to reach totalitarian results. The ambition of total monitoring and modification of our life and behavior is the ambition of *big business* in the tech industry: to appropriate, colonize, commodify, and monetize every last piece of our life and behavior for the sake of profits. As a Silicon Valley developer of learning applications states the corporate mission:

---

[20] Mussolini, B. (1932). *The Doctrine of Fascism.* Verified 24.06.2018: http://www.worldfuturefund.org/wffmaster/Reading/Germany/mussolini.htm

> The goal of everything we do is to change people's actual behavior at scale. When people use our app, we can capture their behavior, identify good and bad behaviors, and develop ways to reward the good and punish the bad. We can test how actionable our cues are for them and how profitable for us.[21]

Regulating behavior in real-time employing gamification and incentives of rewards and punishments as means is a mission shared by the Chinese State Council and Silicon Valley operatives. In China, for the sake of "trustworthy," conformity to norms, social order, and harmony—in Silicon Valley, for the profits. Without regulation, restrictions, and citizen protection in the data economy, a new variant of *Corporate Totalitarianism* may manifest.

Corporate totalitarianism may be defined as the total interpenetration of the social and the *profitable* and the identification of *market* and society. Everything is potentially profitable. There is no value outside of the market—no value but market value. If all aspects of our life are marketized and commodified as raw material for generating profits, the market and the commercial domain of commodities has become all-embracing with nothing existing or having value outside. Quite contrary to libertarians as Rand and Ron Paul's identification of less market regulation with increased individual freedom and self-determination,[22] this sort of market fundamentalism actualized in the age of Big Data may yield totalitarian results. From a citizen perspective, a multiplicity of different actors competing internally in an unregulated market of data and information may result in an information regime not radically different from the centralistic Chinese system. If everything one does is seen, registered, evaluated,

---

[21] Zuboff, S. (2016). "The Secrets of Surveillance Capitalism", *Frankfurter Allgemeine Zeitung* March 5th 2016. Verified January 11th 2017: http://www.faz.net/aktuell/feuilleton/debatten/the-digital-debate/shoshana-zuboff-secrets-of-surveillance-capitalism-14103616.html?printPagedArticle=true

[22] The Technology Revolution: A Campaign for Liberty Manifesto. Verified 24.06.2018:https://www.scribd.com/doc/99220534/The-Technology-Revolution

and get rewarded or sanctioned accordingly, the result is a perverted proxy of an omniscient God judging and making sure everybody *reap as they saw*. *The Big Other* may be at least as powerful and oppressive as Big Brother:

> [The Big Other] is a ubiquitous networked institutional regime that records, modifies, and commodifies everyday experience from toasters to bodies, communication to thought, all with a view to establishing new pathways to monetization and profit. Big Other is the sovereign power of a near future that annihilates the freedom achieved by the rule of law. (Zuboff 2015: 81).

Technological progress is not necessarily progress for humanity—it may be the opposite. "Free stuff" online may just turn out to be extremely expensive, especially if one is not aware of the full price. It may cost us our democracy, our self-determination, and in the end our freedom.

Correction to:
Reality Lost

Correction to:
V. F. Hendricks, M. Vestergaard, *Reality Lost*,
https://doi.org/10.1007/978-3-030-00813-0

The book was inadvertently published without the following
credit line.

*Translated from the Danish by Sara Høyrup / Hoyrup.biz –
English language copy editing by Vincent F. Hendricks*

This has been added to the copyright page of the book.

The updated version of the book can be found at
https://doi.org/10.1007/978-3-030-00813-0

References

Arendt, H. (1951). *The origins of totalitarianism*. New York: Harcourt Brace & Company.

Barkun, M. (Ed.). (2013). *A culture of conspiracy*. Berkeley: University of California Press.

Barnes, J. (1994). *A pack of lies: Towards a sociology of lying*. Cambridge: Cambridge University Press.

Baudrillard, J. (1994). *Simulacra and simulations*. Ann Arbor: University of Michigan Press.

Berger, J., & Milkman, K. L. (2012). What makes online content go viral? *Journal of Marketing Research, 49*(2), 192–205.

Broersma, M., & Graham, T. (2013). Twitter as a new source. *Journalism Practice, 7*(4), 446–464.

Brunnermaier, M., & Schnabel, I. (2017). Bubbles and central banks: Historical perspectives. In M. Bordo et al. (Eds.), *Central banks at a crossroads: What can we learn from history?* (pp. 493–562). Cambridge: Cambridge University Press.

Bryan, D., Lovett J., & Baumgartner, F. (2014). "The diversity of internet media: Utopia or dystopia?" in *Midwest Political Science Association*, Chicago, April 3–6.

Burris, B. (1993). *Technocracy at work*. New York: SUNY Press.

Cambell, W. J. (2016). *Getting it wrong: debunking the greatest myths in American journalism*. Berkeley: University of California Press.

Colombo, A., & Magri, P. (2017). *The age of uncertainty: Global scenarios and Italy*. Novi Ligure: Edizioni Epoké.

Coser, L. A. (1977). *Masters of sociological thought: Ideas in historical and social context*. New York: Harcourt Brace Jovanovich.

Dahlgren, P., & Alvares, C. (2016). Populism, extremism and media: Mapping an uncertain terrain. *European Journal of Communication, 31*(1), 46–57.

© The Author(s) 2019   139
V. F. Hendricks, M. Vestergaard, *Reality Lost*,
https://doi.org/10.1007/978-3-030-00813-0

Dearing, J. W., & Rogers, E. (1996). *Agenda-setting*. Thousands Oaks: Sage.

Druckman, J. (2003). The power of television images: The first Kennedy–Nixon debate revisited. *The Journal of Politics, 65*(2), 559–571.

Edelman, M. (1979). Review of political language. *American Political Science Review, 73*, 840–855.

Elholm, T. (2011). Symbol- og signallovgivning i kriminalpolitisk perspektiv. In Andersson et al. (Eds.), *Festskrift till Per Ole Träskman* (pp. 166–178). Stockholm: Norstedts Juridik.

Esser, F., & Matthes, J. (2013). Mediatization effects on political news, political actors, political decisions, and political audiences. In H. Kriesi, S. Lavanex, F. Esser, J. Matthes, M. Bühlmann, & D. Bochsler (Eds.), *Democracy in the age of globalization and mediatization* (pp. 177–201). Basingstoke: Palgrave Macmillan.

Festinger, L. (1957). *A theory of cognitive dissonance*. Stanford: Stanford University Press.

Flynn, D. J., Nyhan, B., & Reifler, J. (2017). The nature and origins of misperceptions: Understanding false and unsupported beliefs about politics. *Political Psychology, 38*(S1), 127–150.

Foucault, M. (1979). *Discipline and punish: The birth of the prison*. New York: Vintage Books.

Frankfurt, H. (2005). *On bullshit*. Princeton: Princeton University Press.

Freud, S. (1917). *A general introduction to psychoanalysis* (p. 251). Auckland: PDF BooksWorld.

Friedman, M. (1953). *Essays in positive economics*. Chicago: University of Chicago.

Gordon, J. (1997). John Stuart mill and the "marketplace of ideas". *Social Theory and Practice, 23*(2), 235–249.

Haidt, J. (2001). The emotional dog and its rational tail: A social intuitionist approach to moral judgment. *Psychological Review, 108*, 814–834.

Haidt, J. (2012). *The righteous mind: Why good people are divided by politics and religion*. New York: Penguin Group.

Herrera, L. (2015). Citizenship under surveillance: Dealing with the digital age. *International Journal of Middle East Studies, 47*(2), 354–356.

Hendricks, V. F., & Hansen, P. G. (2014/2016). *Infostorms: Why do we "like"? Explaining individual behavior on the social net* (2nd Rev. and expanded edition). New York: Copernicus Books/Springer Nature.

Hindman, M. (2009). *The myth of digital democracy*. Princeton: Princeton University Press.

Hjarvad, S. (2008). *En verden af medier: Medialiseringen af politik, sprog, religion og leg*. Frederiksberg: Forlaget Samfundslitteratur.

Hume, D. (1739). *Treatise on human nature, book II: The passions*. Online, verified 06.12.2017: http://www.earlymoderntexts.com/assets/pdfs/hume1739book2.pdf

Humprecht, E., & Esser, F. (2017). Diversity in online news: On the importance of ownership types and media system types. *Journalism Studies*, online first, pp. 1–23 | Published online: April 10, 2017.

Hurley, M., & Adebayo, J. (2017). Credit scoring in the era of big data. *Yale Journal of Law and Technology, 18*(1), Article 5.

Iyengar, S., & Hahn, K. S. (2009). Red media, blue media: Evidence of ideological selectivity in media use. *Journal of Communication, 59*, 19–39.

Kahneman, D. (1973). *Attention and effort*. Englewood Cliffs: Prentice-Hall Inc..

Kahneman, D. (2011). *Thinking, fast and slow*. New York: Farrar, Straus and Giroux.

Klinger, U., & Svensson, J. (2015). Network media logic: Some conceptual considerations. In *The Routledge companion to social media and politics* (pp. 23–38). New York: Routledge.

Korsgaard, L. (2017). *Den der råber lyver. Mediebrugerens manual til løgnenes tidsalder*. Copenhagen: Lindhardt og Ringhof.

Kriesi, H. et al. (2013). *Democracy in the age of globalization and mediatization*. Springer Nature.

Kuklinski, J. H., Quirk, P. J., Jerit, J., Schwieder, D., & Rich, R. F. (2000). Misinformation and the currency of democratic citizenship. *The Journal of Politics, 62*, 790–816.

Lewandowsky, S., Oberauer, K., & Gignac, G. E. (2013). NASA faked the Moon landing – Therefore, (Climate) science is a hoax. *Psychological Science, 24*, 622–633.

Machiavelli, N. (1999). *The prince*. New York: Penguin.

Manjoo, F. (2008). *True enough: Learning to live in a post-fact society*. New York: Wiley.

Mankiw, N. G. (2010). Spreading the wealth around: Reflections inspired by Joe the Plumber. *National Bureau of Economic Research*, working paper 15846, verified 06.12.2017: http://www.nber.org/papers/w15846.pdf

Marwick, A. (2015). You may know me from YouTube: (Micro-) celebrity in social media. In P. D. Marshall & S. Redmond (Eds.), *A companion to celebrity* (pp. 333–349). New York: Wiley-Blackwell.

Mazzoleni, G., & Schulz, W. (1999). 'Mediatization' of politics: A challenge for democracy? *Journal of Political Communication, 16*(3), 247–261.

McCombs, M. E., & Reynolds, A. (2002). New influence on our pictures of world. In E. Jennings & D. Zilmann (Eds.), *Media effects: Advances in theory and research* (pp. 1–18). Mahwah: Lawrence Erlbaum Associates Publishers.

McCombs, M. E., & Shaw, D. L. (1972). The agenda-setting function of mass media. *The Public Opinion Quarterly, 36*(2), 176–187.

McLuhan, M., & Flore, Q. (1967). *The medium is the message*. New York: Penguin Books.

Mocanua, D., Rossia, L., Zhanga, Q., Karsaib, M., & Quattrociocchi, K. (2015). Collective attention in the age of (mis)information. *Computers in Human Behavior, 51*(Part B), 1198–1204.

Morris, J. S., & Francia, P. L. (2009). Cable news, public opinion, and the 2004 party conventions. *Political Research Quarterly, 62*(3), 345–355.

Müller, J. W. (2016). *What is populism?* Philadelphia: University of Pennsylvania Press.

O'Neil, C. (2016). *Weapons of math destruction: How big data increases inequality and threatens democracy*. New York: Crown.

Packard, V. (1960). *The hidden persuaders*. New York: Penguin.

Peterson, R. L. (2016). *Trading on sentiment: The power of minds over markets*. New York: Wiley.

Postman, N. (1985). *Amusing ourselves to death: Public discourse in the age of show business*. New York: Penguin Books.

Quiggin, J. (2010). *Zombie economics. How dead ideas still walk among us*. Princeton: Princeton University Press.

Samuelson, P. A., & Nordhaus, W. D. (2010). *Economics*. Boston: McGraw-Hill Irwin.

Schmitt, C. (1932). *The concept of the political* (p. 2007). Chicago: University of Chicago Press. Expanded Edition.

Schopenhauer, A. (1966). *The world as will and representation*. New York: Dover Publications.

Shiller, R. (2017, January 7). Narrative economics. Cowles Foundation Discussion Paper, no. 2069, Yale University, presidential address delivered at the 129th annual meeting of the American Economic Association, Chicago.

Siebert, F. S., Peterson, T., & Schramm, S. (1956). *Four theories of the press: The authoritarian, libertarian, social responsibility, and Soviet communist concepts of what the press should be and do*. Chicago: University of Illinois Press.

Simon, H. A. (1971). Designing organizations for an information-rich World. In M. Greenberger (Ed.), *Computers, communications, and the public interest* (pp. 38–52). Baltimore: Johns Hopkins Press.

Skogerbo, E., Bruns, A., Quodling, A., & Ingebretsen, T. (2016). Agenda-setting revisited: Social media and sourcing in mainstream journalism. In A. Bruns, G. S. Enli, E. Skogerbo, A. O. Larsson, & C. Christensen (Eds.), *The Routledge companion to social media and politics* (pp. 104–120). New York: Routledge.

Søe, S. O. (2014). Information, misinformation og disinformation En sprogfilosofisk analyse. *Nordisk Tidsskrift for Informationsvidenskab og Kulturformidling, 3*(1), 21–30.

Sternberg, R., & Sternberg, K. (2012). *Cognitive psychology* (6th ed.). Belmont: Wadsworth.

Taplin, J. (2017). *Move fast and break things: How Facebook, Google, and Amazon cornered culture and undermined democracy*. New York: Little, Brown and Company.

Teixeira, T. S. (2014). *The rising cost of consumer attention: Why should you care, and what you can do about it* (Harvard Business School, Working Paper 14-055), January 17, 2014.

Thesen, G. (2013). Political agenda setting as mediatized politics? Media–politics interactions from a party and issue competition perspective. *The International Journal of Press/Politics, 19*(2), 181–201.

Thorson, E. (2016). Belief echoes: The persistent effects of corrected misinformation. *Political Communication, 33*, 460–480.

Vogel, H. L. (2010). *Financial markets and crashes*. New York: Cambridge University Press.

Voltaire, F. M. (1991). *Candide*. New York: Dover Publications.

Vosoughi, S., Roy, D., & Aral, S. (2018). The spread of true and false news online. *Science, 359*, 1146–1151.

Weber, M. (1949). *On The methodology of the social sciences* (Translated and edited by E. Schils & H. Finch). New York: The Free Press.

Webster, J. G. (2014). *The marketplace of attention*. Cambridge: The MIT Press.

Wu, T. (2016). *The attention merchants*. New York: Knopf.

Zuboff, S. (2015). Big other: Surveillance capitalism and the prospects of an information civilization. *Journal of Information Technology, 30*, 75–89.